ASTRONOMY

D0881514

ABOUT THE AUTHOR

Roger B. Culver received his doctorate degree in astronomy from Ohio State University in 1971. He has taught at Colorado State University since 1966 and currently holds an associate professorship at that institution. He is also a member of the American Astronomical Society.

For the past ten years Dr. Culver has been a guest observer at the Kitt Peak National Observatory, the Perkins Observatory, and the Lowell Observatory. He has published papers in several areas of observational astronomy, and is the author of a laboratory text, *An Introduction to Experimental Astronomy*. He is also co-author of a scientific look at astrology entitled *The Gemini Syndrome*.

Dr. Culver was voted Teacher of the Year by his colleagues at Colorado State in 1977, and is listed in *American Men and Women of Science* and *Outstanding Educators of America*.

ASTRONOMY

Roger B. Culver

BARNES & NOBLE BOOKS

A DIVISION OF HARPER & ROW, PUBLISHERS

New York, Hagerstown, San Francisco, London

To Bonnie, Ted, and Louise
with love

BARNES AND NOBLE BOOKS edition published 1979.

ISBN: 0-06-460158-7

LIBRARY OF CONGRESS CATALOG NUMBER: 78-15832

Designed by Eve Kirch

79 80 81 82 83 10 9 8 7 6 5 4 3 2 1

Contents

Preface

Today, when a gulf still exists between scientists and nonscientists, astronomers occupy an enviable position. Not only do they hold the respect of the scientific community as *bona fide* investigators of the physical world, they also have captured the imagination of the nonscientific community with speculations on such topics as black holes, quasars, and life on other planets. Because these subjects have such appeal, college-level nonscience students seldom, if ever, enter an introductory astronomy course with the same terror with which they approach physics and mathematics courses. The introductory astronomy course thus provides a unique opportunity to render the two-culture gap less formidable.

To meet the needs of students in such courses, a wide variety of astronomy texts have appeared in recent years. Although their common aim is to present a technical subject in an interesting and accessible way, they do vary considerably in emphasis and approach. This *Outline* in astronomy has been designed to be reasonably compatible with all these texts. Thus it should provide students with a comprehensive and compact summary of the course for study, reference, and review. It should also appeal to the general reader who wants an up-to-date survey of astronomy.

The material is arranged largely in the traditional history, instruments, "inside-out" approach. This permits the student to begin with more familiar topics, such as history, and from there jump onto more difficult and unfamiliar ground. The level of mathematics never exceeds that of a good college algebra course, and most of the time it is held to a much lower level. Metric units, with emphasis on those of the CGS system, are employed throughout except in cases such as the 200-inch Mt. Palomar reflector, where the weight of history is simply too great to be ignored.

I am deeply grateful to my wife Bonnie and to my family and friends for their support and encouragement during the course of the project. I also want to express my appreciation to the support staff at Colorado State University, who converted an almost totally indecipherable draft into a presentable manuscript and to David Mack for a number of useful comments. Thanks are also due to the individuals and institutions who gave their kind permission to use the illustrations in this text. I especially wish to thank Jeanne Flagg, editor, Barnes & Noble Books, for her expert assistance. But most of all, I am indebted to the students of Astronomy 100 at Colorado State University who, over the past several years, have provided valuable input from the other side of the gulf.

ROGER B. CULVER

1

The History of Astronomical Thought

No aspect of the physical world has so strongly aroused human curiosity as the heavens. The full moon illuminating a landscape, the planet Venus shining in the evening twilight, a shooting star cutting the darkness, a solar eclipse turning day to night, and a comet trailing across the sky are but a few examples of the awesome celestial beauty that has both fascinated and frightened human beings since prehistoric times.

Although little is known of early observation of the heavens, such observations were almost certainly made with some degree of sophistication. We know, for example, that all the naked-eye planets were recognized as "wanderers" among the fixed stars as far back as written records exist. Thus, the beginnings of the science of astronomy easily predate those of the other natural sciences.

NON-GREEK ANCIENT ASTRONOMY

In ancient times, explanations of celestial objects and phenomena were deeply interwoven with the mythology, legends, and religious beliefs of the various cultures. Thus, the sun was regarded by the Incas and Egyptians as the physical manifestation of the most important of their gods, and the long-tailed comets were universally thought to be divine messengers sent to warn of approaching disasters. Within this mythological framework, the ancient cultures were able to develop an astronomy that dealt with problems of a practical nature. Of particular interest are the origins of timekeeping, celestial navigation, and astrology. Each of these areas required not only that fairly systematic observations be made of the sun, moon, planets, and stars, but also that these observations be used to determine some specific piece of information.

Timekeeping. Perhaps the most practical use of astronomical observations is in timekeeping. The ancients became aware that certain celestial events—the rising and setting of the sun, the cycle of lunar phases, and the apparent motion of the sun among the stars—repeated themselves after fixed intervals of time, and used these events to define the basic units (day, month, and year) with which they reckoned time.

The calendar was developed as an orderly way of relating the day, month, and year. For some civilizations, the calendar served to predict a natural event of crucial importance, such as the flooding of the Nile River. For others, the calendar ensured that feast days devoted to certain gods would be celebrated at exactly the right time, thus avoiding the wrath of, say, the fertility god at planting time. Because the relationships among the day, the year, and the lunar cycle are complicated, the development of an accurate calendar requires detailed astronomical observations of the sun and moon. In some early civilizations, temples were built in such a way that celestial events could be observed and used to reckon the time of the year. For example, the Egyptians aligned some of their temples so that just prior to the onset of the flooding of the Nile River, the bright star Sirius would shine through the entrance at dawn; and the builders of Stonehenge in southwestern England seem to have erected the stones to align with certain significant celestial positions such as midsummer sunrise.

The calendar systems devised by the cultures of the ancient world were many and varied. The Babylonians divided the year into twelve 30-day months and corrected their 360-day year every six years or so by adding a thirteenth 30-day month. The Mayas of Central America had an 18-month year, each month of which had 20 days; at the year's end, five days were added on to correct for the shortage. The early Romans employed a 10-month lunar calendar having alternating 29- and 30-day months. Every three years, a 30-day month was added. By the time of Julius Caesar, a 12-month calendar was in use; however, it was in poor correspondence with the seasons and in great need of adjustment. In 46 B.C., Caesar put into effect a 12-month, 365-day calendar in which no attempt was made to incorporate the lunar cycle and in which every fourth year would be a leap year containing 366 days. The slight discrepancy between the Julian year and the tropical year (11 minutes 14 seconds) eventually became a problem; the vernal equinox, and consequently the date of Easter, was occurring earlier and earlier. This trend was halted in 1582 by the Gregorian reform. Ten days were dropped that year, and thereafter, only century years divisible by 400 were to be leap years, making the calendar accurate to one day in 3323 years. The Gregorian calendar is the one we use today.

Celestial Navigation. Many ancient civilizations were maritime in nature and used the sun, moon, and stars as guideposts in sailing from one point to the other. The people of the Pacific Islands were especially adept at deep-water navigation. The Polynesians devised the sacred calabash, a gourd in which four holes were bored at the same height near the neck. The gourd was then filled with water to the level of the holes, and using the water level as a horizon, altitudes of the stars were measured by sighting through one of the holes over the opposite edge of the gourd. The Caroline Islanders developed a star compass in which 32 "points" were defined by the points on the horizon where prominent stars were observed to rise and set.

Unfortunately, a great deal of information regarding the method and techniques of the ancient celestial navigators has been lost. In some instances, the knowledge was lost when the culture was destroyed, as in the case of the Carthaginians and Phoenicians. In others, the culture had an oral tradition of transmitting knowledge from generation to generation and left very little in the way of permanent records.

Astrology. People have long been fascinated by the possibility of predicting their future destiny and have devised a number of methods of doing so, including palm reading and the use of Tarot cards. One of the oldest of these, astrology, arose about 2500 B.C. in Mesopotamia. For the astrologer, human personality and human affairs are controlled in a predictable way by the positions of the sun, moon, and planets relative to the fixed stars and to each other. Because such prediction requires systematic and accurate observations of the sky, which are much more demanding than those needed for adequate calendar making or celestial navigation, astrology was the driving force behind the development of better astronomical observations among the ancients, especially the Mesopotamian civilizations. Moreover, astrology was often the only means by which someone otherwise interested in the heavens could obtain material support, and in a great many cultures, it was the respected and powerful court astrologer who developed and preserved that culture's store of astronomical knowledge. The relationship between astronomy and astrology was in many respects similar to that which existed between chemistry and alchemy, and like that relationship, was swept away in the scientific revolution of the sixteenth and seventeenth centuries.

GREEK ASTRONOMY

One notable exception to the pragmatism of the ancient cultures was the civilization developed by the Greeks. Perhaps history's greatest thinkers, the Greeks devoted a great deal of time and effort to problems of an abstract nature with little regard for their practical use. Interestingly enough, many of the ideas basic to our modern scientific civilization, such as atomic theory, have their roots in the culture of ancient Greece. Ancient Greek speculations about nature, unlike those of their contemporaries in other lands, were not made idly or even with the idea that the true nature of the universe could ever be fully understood. Instead, their speculations were made within the framework of a kind of philosophical game in which a scheme would be proposed that would describe and explain a certain effect or aspect of nature and at the same time would "save the phenomena," that is, account for all the observed aspects of a given effect.

Pythagoras. The astronomical problem that commanded by far the greatest interest among the Greek thinkers was explaining the motions of the sun, moon, and planets in terms of a single, simple mathematical model. Pythagoras, in the fifth century B.C., explained these motions in terms of a series of concentric crystalline celestial spheres centered on the earth and on which all the observable celestial objects were mounted. The motions of the sun, moon, and planets could be more or less reproduced and predicted by having one sphere for the sun, moon, and each planet and by assigning to each sphere the proper inclination and rotation rate.

Aristotle. In the fourth century B.C. Aristotle also held that the earth was the center of the planetary system. He argued that if the earth were in motion around the sun, then a nearby star would exhibit a *parallax* or apparent angular shift with respect to the more distant background stars if observed from two different positions in the earth's orbit (see Fig. 3.4). Since no parallax could be detected, Aristotle reasoned that either the earth was stationary and was the center of the universe or all the stars were at such great distances that none of them would exhibit a parallax large enough to be detected with the naked eye. We know that the latter explanation is correct, but Aristotle seemingly could not accept the idea of a vast void between Saturn, the last of the then-known planets, and the nearest stars. He thus chose to accept an earth-centered or *geocentric* universe. This concept was challenged by Aristarchus of Samos, who determined that the sun was a larger body than the earth and concluded that the universe was *heliocentric,* or sun-centered. Aristotelian ideas and explanations, however, came to dominate in the latter years of classical antiquity, and Aristarchus's heliocentric model was, for the time being, cast aside.

Hipparchus. Foremost of the Greek astronomers was Hipparchus of Alexandria, whose list of impressive accomplishments during the second century B.C. includes an accurate determination of the moon's distance, cataloguing over 1000 stars according to their positions and apparent magnitudes, and years of observations of the apparent positions of the sun, moon, and planets made with an accuracy that would not be exceeded until the sixteenth century. Hipparchus, also failing to detect a stellar parallax, embraced the geocentric theory and introduced an ingenious refinement to earlier geocentric models in which celestial motions could be represented by *eccentrics,* or off-center circles; *deferents,* or circular orbits centered on the earth; and *epicycles,* or small circular orbits centered on a point moving about the earth along a deferent or eccentric (see Fig. 1.1). With this system, he was able to match closely the lunar and solar motion, but the motions of the planets, with their "backward" loops, or retrograde motion, were sufficiently complicated that Hipparchus was unable to account for them accurately with a simple geometric model.

Ptolemy. The man who attacked the problem of planetary motions with a certain amount of success was Claudius Ptolemy, who, like Hipparchus, worked at Alexandria. Using the observations of Hipparchus as well as his own, Ptolemy was able to account for all the solar, lunar, and planetary motions to an accuracy within the observational error of the data available to him. To accomplish this, Ptolemy used various combinations of angular velocities, as well as the circular orbits and geometric points developed by Hipparchus. He also employed *equants,* or off-center points within eccentrics and deferents

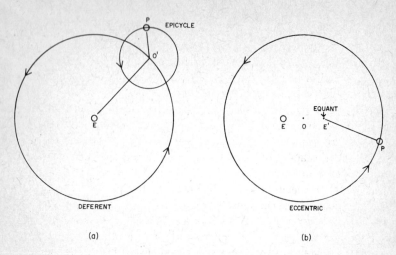

Fig. 1.1 A diagram of the deferent, epicycle, equant, and eccentric. Using various combinations as these figures, Hipparchus and Ptolemy were able to match the motions of the sun, moon, and planets to a fairly high degree of accuracy.

about which an object or an object's epicycle must move at a uniform angular rate. Examples of the success of the Ptolemaic system in explaining planetary motion are illustrated in Figs 1.2 and 1.3.

Ptolemy's system of orbits is summarized along with all of ancient astronomy in his great thirteen-volume compendium on astronomy called the *Almagest*. With his death around A.D. 150, the advance of ancient astronomy was halted and no basic changes in the world model developed by Ptolemy would be forthcoming for thirteen centuries.

MEDIEVAL ASTRONOMY

If Aristotle and Ptolemy could have known that their ideas would be accepted with little or no change for over a millennium, they might have felt complimented, but more likely, they would have been surprised. Such, however, was the outcome of the forces set in motion in Western Europe and the Mediterranean during the last days of the Roman Empire and the years that followed the Fall of Rome.

In general, the Romans lacked the Greeks' enthusiasm for abstract speculation about the nature of the universe and concerned themselves more with practical morality and the philosophy of how one should live so as to achieve peace and prosperity. Indeed, the Romans' admiration for Archimedes of Syracuse, the greatest of the ancient Greek mathematicians, centered not on his abstract mathematical prowess, but rather on the fact that the machines of war he designed during the Roman siege of Syracuse in 212 B.C. almost single-handedly kept the vaunted Roman army at bay for over two and one-half years. The extent to which Roman priorities and concerns permeated the attitudes of the cultures they conquered had, by the second and third centuries A.D., effectively choked off further advances in Greek world models after the death of Ptolemy.

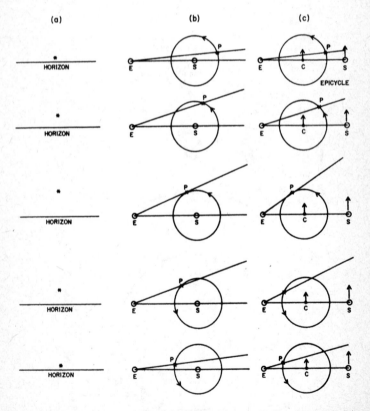

Fig. 1.2 The motion of an interior planet (a) as actually observed, (b) as explained in the heliocentric theory, and (c) as explained in the geocentric theory.

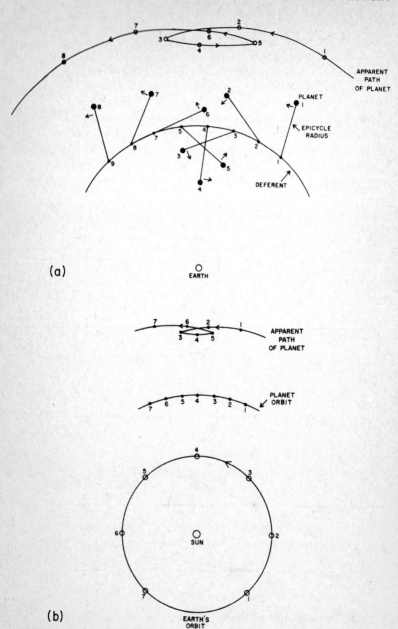

Fig. 1.3 The phenomenon of retrograde motion as explained by (a) the geocentric theory and (b) the heliocentric theory.

After the fall of Rome, Western Europe was fragmented into hundreds of warring feudal states, and the conquest of the Near East and its great learning centers, such as Alexandria, by the Mohammedans in the seventh and eighth centuries, effectively cut off contact between the centers of Hellenistic Greek culture and Western Europe. The rise of Christianity, with its emphasis on a spiritual hereafter and its view of the world as a source of distraction to the efficient pursual of one's salvation, perhaps contributed most heavily to the thousand-year drought in astronomical advance. Particularly devastating was the dogmatism that developed in the Church and required an unquestioning, literal acceptance of the Bible and other approved sources of authority. In the Near East, meanwhile, the Arabs were making slight contributions to astronomy, but basically they acted as caretakers for the store of the Greeks' astronomical knowledge from the time of the Mohammedan rise until the Renaissance.

THE SCIENTIFIC REVOLUTION

By the start of the sixteenth century, the forces that had retarded progress in astronomy since the time of Ptolemy had begun to run their course. The emerging nation-states of Western Europe gradually replaced the multitude of warring feudal states and provided more unified and coherent governments. Moreover, the Christian Church of sixteenth-century Europe was a far more worldly institution than it had been in its earlier days. For example, the Pope, as spiritual leader of all Christendom, also governed the Papal States in Central Italy, participated in military campaigns and political intrigues, and in general, conducted his affairs just like any other head of a temporal state. These and other trends toward a materialistic outlook within the Church, such as fees for indulgences, disturbed a significant part of the Church membership, and the discontent flared into the open in 1517 with the onset of the Protestant Reformation.

Earlier, the Crusades, while failing in their avowed task of regaining the Holy Lands for Christianity in the twelfth and thirteenth centuries, nevertheless had contributed greatly to the reopening of a viable relationship between the Near East and Western Europe. One result of this renewed contact was an increased interest in Greek ideas and writings in the West. As early as the thirteenth century, St. Thomas Aquinas (1225–1274) wrote his *Summa Theologica,* a brilliant amalgamation of Christian dogma and Aristotelian morality that serves as the basis for much of Christian philosophy to this day. Indeed, the ideas of the Greek writers, especially Aristotle, were accorded the same blind acceptance that was normally reserved for articles of the Christian faith. The world model of Ptolemy was particularly revered, not only because it seemed to be in keeping with a literal interpretation of several Biblical passages, but more important, because it placed man, the greatest of God's creatures, at the center of the universe.

Copernicus. The Ptolemaic system, however, was not to withstand the outburst of intellectual activity that characterized the Renaissance. To compensate for the disagreement between the observed planetary motions and what the Ptolemaic system predicted, additional epicycles were added by various workers; by 1540, a total of 79 orbits and epicycles was necessary to describe planetary motions. The task of simplifying the unwieldy system of orbits and epicycles was assumed by Nicolas Copernicus (1473–1543), a Polish mathematician and astronomer. In so doing, Copernicus revived the heliocentric theory of Aristarchus. Unfortunately, he could not completely abandon the belief held for centuries that the paths of celestial objects were combinations of uniform circular motions, and found it necessary to make use of 34 epicycles to account for the motions of the earth, moon, and planets about the sun. Because Copernicus could offer no proof that his system more correctly described the universe than that of Ptolemy, Church authorities were quick to brand the new theory, published in his book, *De Revolutionibus,* in 1543, the year of his death, as nothing more than a convenient model for mathematical calculations. Despite the fact that Copernicus's heliocentric system was only a partially successful model, it nonetheless served to restore interest in the general problem of planetary motions.

Tycho Brahe. The relative ease with which both a heliocentric and a geocentric model could be fit to the planetary-motion data pointed to the inaccuracy of astronomical observations of the era. This situation came to an end in the latter half of the sixteenth century with the rise to prominence of perhaps the greatest naked-eye observer of all time, the Danish astronomer, Tycho Brahe (1546–1601). Gifted with keen eyesight and a facility for designing instruments of great accuracy, Tycho, from his Uraniborg Observatory in Denmark, for over twenty years conducted observations of the motions of the planets that far surpassed anything previously accomplished. Ultimately, Christian IV, King of Denmark, who was Tycho's last patron and sponsor, withdrew his support over costs and personality conflicts, and Tycho moved to Prague in 1597.

Kepler. Tycho soon was joined by a brilliant young mathematician, Johannes Kepler (1571–1630), whose theoretical abilities complemented Tycho's experimental talents. Kepler spent many years trying to formulate a model for the planetary system from Tycho's data. He found that neither a geocentric model nor Tycho's geostatic idea of a system of planets orbiting the sun, which in turn orbited the earth, could be reconciled with the data. Even Copernicus's heliocentric model presented some problems in terms of agreement within the limits of Tycho's observational error. Working principally with Mars, whose orbit was particularly troublesome, Kepler tried to fit various combinations of epicycles and eccentrics to Tycho's data. After many years, he found, to his surprise, that if he represented the planetary paths by ellipses, oval-shaped mathematical figures (see Fig. 4.2), the orbits fit Tycho's data to the degree of accuracy he sought. Thus, Kepler found the first of his planetary laws:

Each planet moves about the sun in an elliptical path with the sun at one focus of the ellipse.

Further study revealed that the planets speeded up nearer the sun and slowed down when farther away. The second law of planetary motion, called the *law of areas,* states:

The planet will move about the sun in such a way that the line joining the sun and the planet will sweep out equal areas in equal periods of time.

Both laws were published in Kepler's work, *The New Astronomy,* in 1609. A decade later, in *The Harmony of the Worlds,* Kepler set down the third and last of his laws of planetary motion, the *harmonic law:*

The squares of the orbital periods of the planets are directly proportional to the cubes of their mean distances from the sun.

These laws will be discussed in greater detail in Chapter 4, but it suffices here to say that the agreement obtained by Kepler between the planetary position data of Tycho and the predictions from these laws finished the geocentric model once and for all.

Galileo. The central drama of the Scientific Revolution, however, was to be played out to the south, in Italy, where yet another of the Renaissance giants, Galileo Galilei (1564–1642), was already on a collision course with Church authorities.

Galileo refused to accept statements solely on the basis that they were set down by writers of accepted authority, and his experiments on bodies in motion disproved many of the Greek ideas on the subject. In 1609, Galileo, hearing of the invention of the telescope in Holland, constructed several of his own and began systematic observations of various celestial objects. His first results were published in 1610 in *Sidereus Nuncius*. He found that the planets Mercury and Venus exhibited a series of phases just as does the moon. He also observed spots on the sun, mountains and craters on the moon, and four bodies revolving around Jupiter. Many of these discoveries challenged the widely held beliefs of the day and pointed to the correctness of the Copernican theory, which Galileo had embraced since the 1590s.

Finally, in 1616 the Church issued a decree which stated that the heliocentric theory was false and absurd and was not to be defended or believed. In spite of this decree, Galileo secured permission to publish his ideas on the heliocentric theory ostensibly for discussion purposes only, and the resulting work, *Dialogue on the Great World Systems*, was published in 1632. Far from being regarded as simply a discussion of world models, the *Dialogue* was received as a devastating assault on the geocentric theory. The Church, smarting from its losses during the first century of the Protestant Reformation and highly sensitive to any departure from orthodox belief, reacted strongly to the impact of the *Dialogue* and brought Galileo before the Roman Inquisition for holding beliefs contrary to the Sacred Scriptures. It forced him to retract his belief in the heliocentric theory, then placed him under house arrest for the remaining ten years of his life. Galileo had, however, established the cornerstone of modern scientific philosophy that no statement regarding nature should be accepted unless it can be verified through observation and experimentation.

Newton. Within a half-century of the death of Galileo, the English physicist and mathematician Isaac Newton (1643–1727), in his fundamental work, the *Philosophica Naturalis Principia Mathematica*, published in 1687, demonstrated that the laws of motion developed by Galileo, Kepler's laws of planetary motion, and other motion phenomena as well were different manifestations of the same fundamental laws governing all motion in the universe (see Chapter 4). The importance of the *Principia* lay in the fact that for the first time in recorded history a single theory could be applied to a wide range of phenomena extending from the swinging pendulum to the motions of the planets.

A dramatic astronomical demonstration of the power of Newtonian theory came in 1846 when two mathematicians, Leverrier of France and Adams of England, independently predicted the existence of an unknown planet beyond the orbit of the last of the then-known planets, Uranus. The calculations were done on the basis of disagreements between the predicted motion of Uranus and its observed motion and were so unerring that the postulated planet Neptune was observed almost exactly in its predicted spot on the very first night of the search.

With the death of Newton, the scientific revolution came to a close. In little more than a century the scientific thinkers of the Renaissance had fashioned the empirical method of knowledge gathering that we now call the scientific method. Combining the Greek idea of "saving the phenomena" with the experimental approach of the Renaissance, the scientific method has continued into the twentieth century as the most viable technique for gaining an understanding of the physical universe.

The Scientific Revolution also marks the parting of the ways between astronomer and astrologer. After the seventeenth century, no longer would the terms astrologer and astronomer be interchangeable. Through the use of the scientific method, the astronomer uncovered a universe far different from that perceived even a few decades earlier. On the other hand, the astrologer made little or no attempt to verify astrological principles by experimental means. As a result, the traditional beliefs of the astrologer, which underlie the casting of horoscopes, have remained virtually unchanged since antiquity. Investigations by modern experimental science have disproved many, if not all, of these traditional ideas. Statistical studies reveal, for example, that there is no correlation between the so-called sun sign (the region of the sky in which the sun is located at the exact time of birth) and the personality traits, physical characteristics, and occupations that astrologers attribute to these signs. When confronted with a prediction that does not come true, the scientist is duty-bound to reexamine the principles and assumptions on which the prediction was based and bring them in line with the experimental results. The astrologers, meanwhile, consider every possibility *but* the invalidity of astrological principles in trying to account for the discrepancies between their predictions and the actual outcomes.

MODERN ASTRONOMY

Essential to the scientific method that emerged from the seventeenth century is the ability of the investigator to make accurate, reproducible measurements of the phenomenon under scrutiny. Galileo's use of the telescope as an astronomical instrument provided the necessary breakthrough for the science of astronomy. For example, the discovery of the light-time effect by Roemer in 1675, the discovery of the aberration of starlight by Bradley in 1729, and the first measurements of the long-sought stellar parallaxes in 1838 by Struve and Bessel, all of which confirmed the revolution of the earth about the sun, would not have been possible without the steadily improving telescopes in the two and one-half centuries after Galileo. During the nineteenth century astronomers came to rely on other auxiliary instruments, in particular, the photographic plate, which allowed the astronomer to make objective, permanent records of telescopic observations.

The ever-increasing precision with which scientific observations were made after the time of Newton led to the discovery late in the nineteenth century of several astronomical phenomena that could not be accounted for in terms of the classical scientific theory of the 1800s. For example, the mechanism by which the sun and distant stars were able to generate their tremendous energy output over millions of years could not be explained by classical physics; nor could classical theory account for either the anomalous behavior of the motions of the planet Mercury as it wheeled about the sun or the mysterious dark lines that appeared whenever the light from the sun or a star was broken into its component colors by means of a spectrograph.

Early in the twentieth century, scientists sought to develop more extensive physical theories that could account for these as well as other unexplainable phenomena observed in other branches of natural science. Out of these efforts came relativity theory and quantum theory. These physical theories have not only explained the "unsaved" phenomena of the last century but have helped to account for the experimental results obtained in the first half of the present century.

This century has also seen the discovery of a wide variety of forms of radiant energy or electromagnetic radiation, which are similar to visible light but which cannot be detected by the unaided human eye (see Chapter 3). Although most of this invisible energy coming to us from distant celestial objects is blocked by the earth's atmosphere, some of it does reach the earth's surface and can be studied with appropriate detectors and recorders. Observations of this type, particularly in the so-called radio region, have led to a number of important astronomical discoveries, the most noteworthy of which are the pulsars (see Chapter 11) and the quasars (see Chapter 15).

The launching of the first artificial satellite in October 1957 added yet another dimension to modern observational astronomy; astronomers can now orbit instruments and detectors above the earth's atmosphere, thereby gaining access to the entire range of electromagnetic radiation. The space age has also seen the development of space probes, instrument packages designed to secure close-up views of other planets with a resolution impossible to obtain with earth-based telescopes, as well as manned space programs such as Apollo, Skylab, and the space shuttle.

The instrumentation developed in this century has uncovered a universe of a vastness and complexity that was unimaginable only a few decades ago. It has also produced tantalizing indications that once more there exist phenomena and observational results that cannot be accounted for by current physical theories.

Unfortunately, there are also some aspects of the physical universe that have proved somewhat elusive for the scientific method. The controversy over the true nature of the UFOs or unidentified flying objects has raged unresolved for over three decades because the observational data for these phenomena come to us almost totally from unexpected visual sightings made by untrained observers. The true nature of the UFOs is thus deeply buried in the poor quality of the observational data, and there seems to be no prospect for an improvement in the situation anytime soon.

REVIEW QUESTIONS

1. Compare and contrast the astronomy of the ancient Greeks with that of other ancient civilizations.
2. Briefly discuss the astronomical contributions of Ptolemy and Hipparchus.
3. Why was no progress made in astronomy from the time of Ptolemy to the time of Copernicus? Why did this lack of progress end?
4. Compare the science that emerged after Newton with that of the Greek scholars.
5. What changes have occurred in the science of astronomy in this century?

2

Astronomical Instrumentation

Virtually all the knowledge of the universe that astronomers have gathered depends in whole or in part on analyses of the electromagnetic radiation received from celestial objects. Prior to the Renaissance, astronomers could study this light only with the naked eye. Galileo's discoveries with his simple telescopes marked the first success in extending human vision beyond the limits of the unaided eye. Since that time the telescope has evolved into a powerful astronomical tool, and the modern astronomer almost never analyzes light from a distant celestial object that has not first been enhanced by means of some sort of telescopic device. To make a detailed analysis of the radiation received by the telescope, astronomers employ a number of auxiliary devices.

THE NATURE OF LIGHT

Light cannot be explained in terms of a single, simple model. In some experiments, such as those on interference phenomena, the observed behavior of light rays can most conveniently be explained by regarding light as a transverse wave motion, similar to the series of concentric ripples generated by dropping a stone into a quiet pond. In this description of light, the distance between successive ripple crests

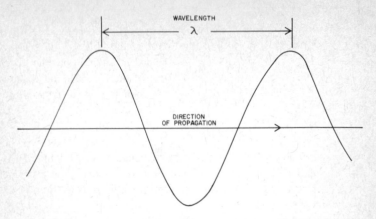

Fig. 2.1 Light as a wave motion.

is called the wavelength λ (see Fig. 2.1). The number of ripple crests or cycles that pass a fixed point per unit time is referred to as the frequency ν of the wave motion, and the speed c at which light waves of any wavelength propagate through space is roughly 300,000 km/sec.

In other experiments, such as those on the photoelectric effect, light behaves as if it were a massless particle possessing particle-like properties such as linear momentum and kinetic energy which are inversely proportional to the wavelength of the light. In this model, the momentum of the photon is equal to h/λ and its kinetic energy is equal to hc/λ, where h is Planck's constant. The speed of the photon in free space is 300,000 km/sec, the same speed as light waves. The actual mathematical description of light is, of course, far more complicated than either of these simple models, but they are nevertheless quite useful in providing an elementary description of the light-related phenomena that are to be discussed.

The Laws of Reflection and Refraction. If unimpeded, light rays will travel in straight-line paths from source to observer. There are a few physical situations in nature, however, in which the paths of light rays are deflected. Two of these deflection phenomena, reflection and refraction, are widely employed in the construction of telescopes. The law of *reflection* tells us that if a beam of light strikes an appropriate surface, it will tend to bounce off or be reflected from that surface in such a way that the angle of incidence and the angle of reflection are equal (see Fig. 2.2a). The phenomenon of *refraction* occurs when a light beam moving in one medium strikes an interface with a second medium at an angle (Fig. 2.2b). Instead of continuing in its original path as it crosses the interface, the light beam will be bent or refracted in a manner that depends on the properties of the materials on either side of the interface. Mathematically this bending is described by *Snell's law:*

$$n \sin \theta_i = n' \sin \theta_r$$

where θ_i and θ_r are the incident and refracted angles, respectively. The quantities n and n' are the so-called indices of refraction of the respective media. The values of n and n' vary from medium to medium and range from 1.0 for a vacuum up to 2.4 for a diamond.

Image Formation. In nature, the images produced by the phenomena of reflection and refraction are distorted, for example, the view of a river bottom through turbulent, swiftly flowing water. However, by shaping a refracting or reflecting surface such as a lens or mirror in just the proper fashion, it is possible to bring a set of incident, parallel light rays from a distant point source to a sharp focus, as shown in Fig. 2.3. The point where the image is formed is called the *focus* or *focal point,* and the distance between the lens or mirror and the focal

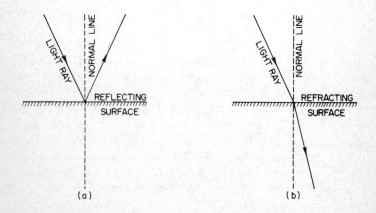

Fig. 2.2 A graphical illustration of light being deflected by (a) reflection and (b) refraction. The dashed "normal" line is perpendicular to the surface at the ray's contact point.

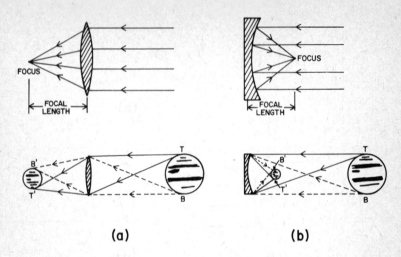

Fig. 2.3 Image formation by (a) a simple thin lens and (b) a concave mirror for distant point sources (top diagrams) and for distant extended sources (bottom diagrams).

point is the *focal length* of the lens or mirror. If an extended source such as a nebula or star cluster is observed, the image is formed on a plane called the *focal plane,* as shown in Fig. 2.3. Each light ray conveys the entire image, and the effect of covering part of the mirror or lens will be only to cut down on the total amount of light forming the image.

Aberrations. All optical systems are plagued with certain inherent imperfections which produce some distortion of the final image. These aberrations are illustrated in Fig. 2.4.

CHROMATIC ABERRATION. Chromatic aberration arises out of the fact that a given lens medium will not refract all colors of light in the same manner. For a thin lens, the focal point for blue light is shorter than that for red light. As long as the incoming light is of a single color (monochromatic), there is no chromatic aberration. If, however, the incoming light consists of several colors, as does sunlight, then color fringes will appear in the image.

SPHERICAL ABERRATION. In a spherical lens or mirror, the outer rays of light focus at a point different from the focal point of the light rays near the center. Such a distortion is called spherical aberration.

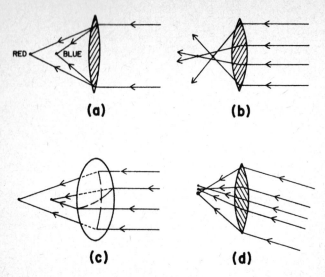

Fig. 2.4 Some common lens aberrations: (a) chromatic aberration; (b) spherical aberration; (c) astigmatism; (d) coma. Because of these aberrations, all optical systems possess a certain amount of image distortion.

ASTIGMATISM. In some optical surfaces, the rays striking the lens in different planes are not focused at the same spot, giving rise to the lens aberration known as astigmatism.

COMA. Some optical systems bring light rays to focus on the focal plane, but not at a single focus point. Images produced by such systems have small comet-like tails pointing radially outward from the center of the field and are said to be affected by coma.

TELESCOPES

There are two basic types of telescopes, refractors and reflectors. Each employs a different method of light gathering and image formation.

The Refracting Telescope. The refracting telescope employs a lens called the *objective* at the front of a tube as the light-gathering and image-forming element (see Fig. 2.5a). The image is formed at the rear of the tube and can be inspected visually by means of a second, smaller eyepiece lens or recorded photographically by placing a photographic plate in the focal plane of the objective lens.

The Reflecting Telescope. In a reflecting telescope a concave mirror serves as the light-gathering and image-forming element. In this case, however, the image formed by the mirror lies between the object being observed and the mirror, thus requiring the use of secondary mirrors to locate the final image at a position convenient for observation.

THE PRIME FOCUS REFLECTOR. The simplest design for a reflecting telescope is the prime focus reflector shown in Fig. 2.5b. In this system no secondary mirrors are used and the image is formed in the middle of the tube. In all but the very largest telescopes, the prime focus is used solely for photographic work. If the image is to be examined visually or by nonphotographic means, the light must be diverted outside the tube before the rays come to focus.

THE NEWTONIAN REFLECTOR. In the Newtonian reflector, the light is diverted by means of a flat mirror mounted at a 45° angle as shown in Fig. 2.5d. The light is thus brought to focus outside the tube. Because of its relative simplicity and convenience of viewing, the Newtonian design is employed most extensively in smaller reflectors.

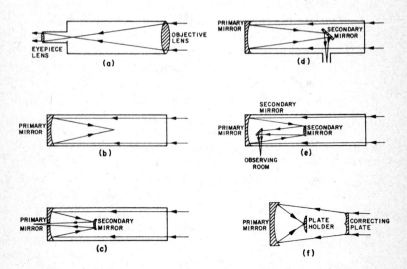

Fig. 2.5 The basic optical systems for astronomical telescopes: (a) the simple refractor; (b) the prime focus reflector; (c) the Cassegrain reflector; (d) the Newtonian reflector; (e) the Coudé reflector; (f) the Schmidt optical system.

THE CASSEGRAIN REFLECTOR. A second method of light diversion is performed by mounting a small convex mirror at the center of the telescope tube. This mirror reflects the light from the primary mirror back down the tube. A small hole in the primary mirror allows the light rays to come to focus at a point behind the telescope tube (see Fig. 2.5c). The Cassegrain reflector allows the astronomer to work at the base of the telescope and is widely used for smaller observatory instruments.

THE COUDÉ REFLECTOR. The Coudé reflector is shown in Fig. 2.5e. As in the Cassegrain reflector, a convex mirror reflects the light from the primary mirror back down the tube. However, the light is then reflected by a flat, diagonal mirror to an observing room. Since the room is stationary and the temperature and humidity can be carefully controlled, it is possible to employ heavy, sensitive equipment at the Coudé focus that could not be used in an open dome.

A Comparison of Reflectors and Refractors. Because reflectors and refractors employ different methods of light-gathering and image formation, each type of telescope has its own set of advantages and disadvantages.

Refractors can be constructed so as to minimize image distortion provided that the incident light is monochromatic. Thus, the refractor finds wide use in planetary and lunar astronomy as well as in the field of astrometry, which deals with the accurate measurement of stellar positions. However, refractors are troubled by chromatic aberration and require lenses of flawless glass, shaped on two sides, which can be supported only about their outer edges. These requirements become prohibitive if a large-aperture refractor is to be constructed, since it is exceedingly difficult to cast and shape a large, flawless piece of glass. Such a lens would also tend to sag slightly if supported only around its rim, thus producing a small distortion of its optical suface. The largest refractor in the world is the 40-inch refractor at the Yerkes Observatory in Williams Bay, Wisconsin, shown in Fig. 2.6.

A reflector, on the other hand, can be constructed so that the primary mirror is supported not only along its rim, but from the back as well without loss of the telescope's ability to gather and focus light. In addition, a mirror needs to be shaped only on one surface. Thus, a reflector does not require a flawless piece of glass and can be made much larger than a refractor. The largest reflector in the world is the 6-meter (236-inch) telescope in the Caucasus Mountains in the southern part of the Soviet Union. The 200-inch reflector of the Mount Palomar Observatory is shown in Fig. 2.7.

Fig. 2.6 The 40-inch refractor of the Yerkes Observatory. (Yerkes Observatory photograph.)

Reflectors are able to focus multicolored light beams, but are not able to focus precisely rays of light that are not near the center of view. Because of their size and lack of chromatic aberration, reflectors are used in spectroscopy and photometry problems where color-corrected images and light gathering are extremely important.

Most large reflectors have a set of secondary mirrors that allow the observer to change from one type of reflecting telescope to another. This, of course, makes the reflector a much more versatile instrument than the refractor, in which the optical design must remain largely unchanged.

The Schmidt Telescope. The Schmidt telescope was devised to combine the advantages of a reflector and a refractor. In the Schmidt optical system, light passes through a thin correcting lens at the front of the telescope and is bent at a slight angle. The light then strikes a primary mirror and is reflected to focus on a curved focal plane as shown in Fig. 2.5f. The Schmidt telescope is thus designed to produce the sharp images of the refractor over a wide field and yet retain the color-corrected images of the reflector. The Schmidt design also allows the telescope to function as an extremely fast camera and is widely used in photographic survey-type problems in which large

Fig. 2.7 The 200-inch reflector of the Mount Palomar Observatory, the largest reflecting telescope in the United States. (Courtesy of the Hale Observatories.)

areas of the sky must be photographed accurately over short periods of time. The most famous of these surveys is perhaps the Palomar Sky Survey, which was conducted in 1950 with the Mount Palomar 48-inch Schmidt telescope illustrated in Fig. 2.8.

Some Telescope Properties. There are several aspects of the image-forming properties of the telescope that are of considerable interest to the astronomer, because these properties determine the limits of usefulness for a given telescope.

LIGHT-GATHERING POWER. One of the most important aspects of any telescope is its ability to serve as a light-gathering device. The amount of light that a telescope can collect and focus is directly proportional to the area of the aperture. Since most telescopes have a circular aperture, whose area is equal to πr^2, where r is the aperture radius, the light-gathering power of a telescope is proportional to both the square of the radius of the aperture and the square of its diameter.

Thus a telescope 20 cm in diameter can collect sixteen times as much light as a 5-cm telescope.

IMAGE BRIGHTNESS. Of crucial importance in recording and viewing an image in the telescope is the brightness B of the image or the amount of energy per unit area concentrated at the focus point. For a point source such as a star, the image brightness is simply proportional to the square of the aperture diameter d, or

$$B \text{ (point source)} = \text{constant} \times d^2$$

For an extended image such as a nebula or star cluster, the image brightness depends not only on the square of the aperture diameter d but also on the focal length f of the lens or mirror as follows:

$$B \text{ (extended source)} = \text{constant} \left(\frac{d}{f}\right)^2$$

The quantity (d/f), which determines the value of B for an extended source, is called the *focal ratio, f ratio,* or *speed* of the objective lens or mirror.

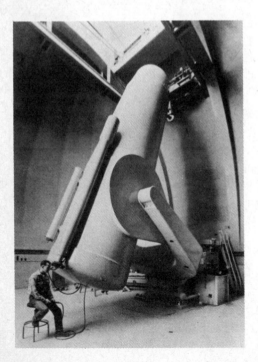

Fig. 2.8 The 48-inch Schmidt telescope of the Mount Palomar Observatory. (Courtesy of the Hale Observatories.)

RESOLVING POWER. Because of the wave nature of light rays, a point source of light is never imaged as a true point in a telescope but rather as a minute spot surrounded by a series of faint concentric rings. This central spot is called a diffraction disk, and its diameter depends on the wavelength of the light being observed and the size of the telescope aperture. Any detail that is smaller than this diffraction disk cannot be resolved or observed with that telescope. The ability of a telescope to resolve fine detail is called its resolving power and is usually expressed as the smallest angular diameter that can be detected with the telescope. This minimum resolvable angle is given by the relation

$$\alpha = \text{constant}\left(\frac{\lambda}{d}\right)$$

where λ is the wavelength of the light being observed and d is the diameter of the telescope aperture. The value of the constant depends on the units chosen for α, λ, and d. If λ is chosen near the center of the visible region, and d is measured in inches, then the above relation reduces to the *Dawes limit formula:*

$$\alpha = \frac{4.56}{d} \text{ (arcseconds)}$$

In reality the effects of atmospheric turbulence are a much more serious hindrance to a telescope's ability to detect detail than is the phenomenon of diffraction.

PLATE SCALE. If the telescope is to be used for photographic work, the scale of the image, or plate scale, is of importance to the astronomer. The plate scale S is usually expressed in arcseconds and is related to the focal length of the telescope by the formula $S = 206.3/f$, where f is the focal length of the objective in meters. Thus the Mount Palomar 200-inch telescope, with a focal length of some 16.8 meters, has a plate scale of 12.3 arcseconds/mm. Most astronomical telescopes have plate scales in the range of 10 to 200 arcseconds/mm.

MAGNIFYING POWER. The ability of a telescope to enlarge an image is called its magnifying power. From the laws of geometric optics, the magnifying power or magnification is equal to the focal length of the telescope objective divided by the focal length of its eyepiece. Thus, an astronomer can alter the magnification of a telescope simply by changing eyepieces. Most large telescopes have a set of eyepieces that can be interchanged for viewing different types of objects under various sky conditions.

TELESCOPIC MOUNTINGS

The telescope must be capable of turning to all parts of the sky. There are two mounting designs, the equatorial mounting and the alt-azimuth mounting.

Equatorial Mounting. The equatorial mounting is by far the most commonly used type of mounting in the larger telescopes. Here one axis of rotation, called the polar axis, is aligned parallel to the earth's axis of rotation as shown in Fig. 2.9a, whereas the other axis is perpendicular to it. Such an alignment allows the telescope to be turned directly in right ascension and declination (see Chapter 3). A small

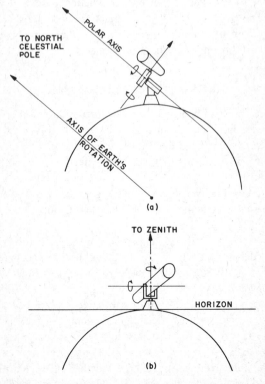

Fig. 2.9 The two basic telescope mountings. The equatorial mounting (a) has its axes aligned perpendicular and parallel to the earth's axis of rotation, whereas the alt-azimuth mounting (b) has its axes aligned perpendicular and parallel to the observer's horizon.

motor attached to the polar axis can be adjusted so as to drive the telescope at just the right rate to compensate for the apparent diurnal motion of the stars due to the rotation of the earth on its axis. By using a drive, astronomers can make long-exposure photographs or other time-consuming observations of a given object without having constantly to adjust the alignment of the telescope tube.

Alt-Azimuth Mounting. In the alt-azimuth mounting, illustrated in Fig. 2.9b, one axis of rotation is aligned perpendicular to the celestial horizon and the other parallel to it. This type of mounting is commonly used for smaller instruments because of the comparative ease with which the mounting's axes of rotation can be aligned. In recent years alt-azimuth mountings have been used for a number of radio telescopes in order to cut down on construction costs and alignment problems. The tracking of celestial objects is accomplished in these large alt-azimuth mountings by rotating *both* the horizontal and vertical axes of the mounting at the proper rates.

"NONVISUAL" TELESCOPES

As we shall see in Chapter 3, there is much radiation that cannot be detected by the human eye. This radiation can, however, be collected and imaged like visible light and thus be made to yield a great deal of information concerning the nature of the universe. The earth's atmosphere absorbs most of this incoming radiation, but certain wavelengths such as radio waves are able to get through. The remaining wavelengths must be studied by satellites orbiting above the earth's atmosphere.

Radio Telescopes. In 1931, K. Jansky of the Bell Telephone Laboratories made the amazing discovery that celestial objects emit radio radiation that can be detected at the earth's surface. Since that time, astronomers have developed huge dish-shaped radio telescopes (Fig. 2.10) that operate in a fashion similar to reflecting telescopes. The incoming radio waves are reflected off a parabolic-shaped metallic antenna and collected at a focal point where the "image" is then examined electronically. Radio waves can be detected at any time of the day or night and even during cloudy weather. Thus, the radio astronomer can operate over a much wider range of sky conditions than can the optical astronomer. Unfortunately, the radio astronomer is also operating at much longer wavelengths than is the optical astronomer, and as a result, the resolving power of radio telescopes is compara-

Fig. 2.10 The 300-foot parabolic dish of the National Radio Astronomy Observatory at Green Bank, West Virginia (National Radio Astronomy Observatory photograph.)

tively low. As an example, the theoretical value of the minimum resolvable angle of the Mount Palomar 200-inch telescope is about 0.023 arcsecond. In order to achieve this same minimum resolvable angle, a radio astronomer operating at a radio wavelength of 21 cm would have to construct a radio telescope having a diameter of roughly 1300 miles! To alleviate this problem, radio astronomers have assembled vast networks of radio telescopes having baselines extending for hundreds of miles. Such networks can, by electronic means, yield the equivalent resolution of a single dish having a diameter equal to the baseline's length.

Satellite Telescopes. In recent years astronomers have orbited various radiation detection devices above the earth's atmosphere in order to observe those wavelengths of radiation that are absorbed by the earth's atmosphere. Although these detectors do not as yet have the sophistication of ground-based instruments, they have nevertheless contributed a great deal to astronomical knowledge.

AUXILIARY INSTRUMENTS

After telescopes have done their job of resolving, magnifying, and gathering light rays, astronomers face the task of extracting as much information as possible from this radiation. In this analysis they make use of a number of auxiliary instruments.

The Camera. The human eye was the primary auxiliary instrument in astronomy well into the nineteenth century, over two centuries after the invention of the telescope. Gradually it was replaced in astronomical work by the more reliable camera. The camera is very similar to the human eye in its operation (see Fig. 2.11a,b). In both, the focus and aperture size can be controlled as the light to be observed passes through the lens. Instead of striking the retina, however, the light in a camera strikes a photosensitive surface called a film emulsion, which

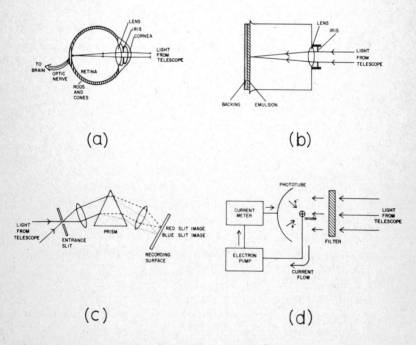

Fig. 2.11 Some basic auxiliary instruments of the astronomer: (a) the human eye; (b) the camera; (c) the spectroscope; (d) the photoelectric photometer.

can then be developed chemically to bring out the image that the camera lens "saw." The camera thus has the advantage of providing an objective, permanent record of a particular observation.

Equally important for the astronomer is the camera's ability to act as a light-collecting device. Unlike the human eye, which responds only to the light that is instantaneously incident upon it, the camera can store light impulses over long periods of time. Thus a nebula or galaxy that appears as only a dim hazy patch of light when observed through a telescope with the human eye may reveal a multitude of intricate detail when photographed for several hours through the same telescope.

The Spectroscope. The spectroscope is a device used by astronomers to break up a beam of light from a star into its component wavelengths or colors. This can be accomplished in a number of ways, including the use of prisms and diffraction gratings. In the simple prism spectrograph, such as that shown in Fig. 2.11c, a light beam is first passed through an entrance slit and then imaged onto the prism by means of a collimating lens. Because the prism bends each wavelength of light differently, the slit image is broken up into a multitude of component slit images each of which has its own unique color or wavelength. A second lens then focuses each of these images onto some sort of recording surface, usually a photographic plate or a scanning phototube. The resulting pattern of light formed at the light-collection surface is a rectangular-shaped series of the component slit images arranged in order of increasing wavelengths. This pattern of light is called a spectrum. Its dispersion, or scale, in angstroms per millimeter (Å/mm), can be adjusted to suit the needs of the astronomer.

The Photoelectric Photometer. An auxiliary instrument that has come into widespread use in recent years is the photoelectric photometer, a device used to measure with great accuracy the brightnesses of celestial objects. A diagram of this instrument is shown in Fig. 2.11d. Light collected by the telescope passes through a color filter and then strikes the surface of a phototube. The material of which the phototube surface is made has the property of liberating electrons (e^-) when struck by light. This is called the photoelectric effect. A positively charged anode near the phototube surface attracts the electrons, which can then be collected and driven around to the phototube by means of an electron pump or voltage supply. As the electron current proceeds to the phototube, its magnitude is indicated by a current-measuring device. A current flows through the circuit only as long as light shines on the phototube surface; moreover, the magnitude of the current varies directly as the intensity of the light. Thus, by measuring the current flowing through the circuit, an astronomer can measure precisely the intensity of the light striking the phototube and hence the brightness of the object being observed. The filter allows the astronomer to measure the object's intensity in various colors of light. Some photometers employ twelve or more such filters.

A wealth of astronomical instrumentation has been developed based on various aspects of the photoelectric effect. These devices include image tubes that photoelectrically intensify the image formed by the telescope, and vidicon or television tubes that record the telescopic image on videotape for analysis by computer techniques.

REVIEW QUESTIONS

1. What are the simple models used to describe light rays?
2. How is it possible to deflect light rays from a straight-line path?
3. Explain how a refracting telescope operates.
4. Explain how a reflecting telescope operates.
5. Compare and contrast the various designs of reflecting telescopes.
6. What is a Schmidt telescope? Compare and contrast the Schmidt telescope with both a reflecting telescope and a refracting telescope.
7. What is an equatorial mounting? Why is it generally used in mounting observatory telescopes?
8. What is an alt-azimuth mounting?
9. If the human eye has a diameter of $^1/_5$ inch, how much more light than the human eye can the Mount Palomar 200-inch telescope gather? *Ans.: =* 1,000,000 times more light.
10. How do resolving power and magnification differ?
11. State two reasons why astronomers would want to build a telescope with as large an aperture as possible.
12. State an advantage and a disadvantage associated with observing celestial objects with (a) a radio telescope and (b) a visual telescope.
13. Describe the various auxiliary instruments used by astronomers. What is the purpose of each?

3

Some Principles of Observational Astronomy

Astronomy is perhaps the most challenging of the observational sciences. Unlike their counterparts in other scientific fields, astronomers cannot experiment and study the objects that command their interest under carefully controlled laboratory conditions, but must analyze the light from objects billions of miles away. In performing this analysis, astronomers assume that the laws which govern physical phenomena here on earth are also at work in distant celestial objects.

ASTROMETRY

Astrometry is that branch of observational astronomy which deals with the measurement of celestial positions. An object's general position in the sky can be specified by making use of a system of 88 sky zones called constellations that have been set up by astronomers. For more precise positions, an astronomer can make use of one of several spherical coordinate systems.

Constellations. Originally, the constellations were star groups chosen by various cultures to represent the heavenly outlines of the heroes, heroines, beasts, and objects prominent in their folklore and mythology. In some constellations, such as Orion, the Hunter, and Scorpius, the Scorpion, the pattern of the stars indeed bears a likeness to its

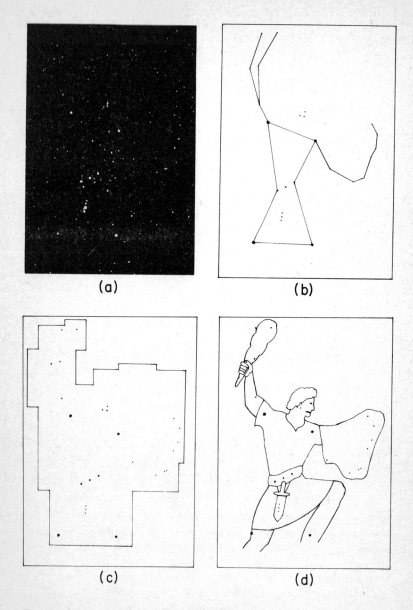

Fig. 3.1 The constellation of Orion: (a) photographic view, (b) modern outline, (c) formal astronomical boundaries, and (d) mythological view. (Roger B. Culver, the Colorado State University Observatory.)

namesake (see Fig. 3.1). In most instances, however, the resemblance is difficult to see.

Although virtually every culture the world over has had its own set of constellations, most of the constellations we know today have come from the Greeks and Babylonians. In more modern times, various countries have had favorite constellations; German star maps, for example, listed a constellation called "Frederick's Glory," the scepter of the famous Prussian ruler, Frederick the Great. Aside from their historical and mythological interest, however, the constellations serve the modern astronomer as large-scale divisions of the night sky. In 1928, the International Astronomical Union (IAU) met to decide what constellations would thenceforth be recognized by the astronomical community and to define the boundaries of these regions. After considerable discussion, a total of 88 constellations and their corresponding regions of the sky were set up by the IAU (see Appendix 6).

Spherical Coordinate Systems. To indicate the more precise celestial positions required for telescopic observations, astronomers employ one of a variety of coordinate systems similar to the latitude-longitude coordinate system used by geographers in specifying positions on the surface of the earth.

THE LATITUDE-LONGITUDE SYSTEM. In the latitude-longitude coordinate system, the earth is regarded as an idealized sphere spinning about an *axis of rotation*. The points of intersection between the earth's surface and this axis of rotation are referred to as the *geographic north pole* and the *geographic south pole*. Exactly halfway between these two points lies a *great circle*, that is, one whose center is at the sphere's center, that divides the earth into equal parts and is hence called the *equator*. The *latitude* of a point on the surface of the earth is then defined as the shortest angular distance between that point and the equator as viewed from the center of the earth. The latitude thus ranges from 90°N at the earth's north geographic pole, through 0° at the earth's equator, to 90°S at the earth's south geographic pole. *Meridians* are great semicircles that have as their endpoints the earth's geographic poles; they are thus at right angles to the equator. One of these meridians, the meridian that passes through Greenwich, England, is designated the *prime meridian* or the meridian from which the longitude of a point is measured. The *longitude* of a point is defined as the shortest angle between the prime meridian and the meridian containing the object and ranges from 0° at the prime meridian to 180° east or 180° west (see Fig. 3.2). Using only these two coordi-

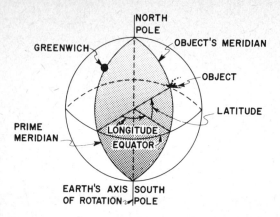

Fig. 3.2 The latitude-longitude system. Any point on the earth's surface can be specified by a pair of angles called latitude and longitude.

nates, geographers are able to determine positions on the surface of the earth to whatever accuracy their instrumentation will allow.

THE CELESTIAL SPHERE. In developing a celestial coordinate system, astronomers have adopted the concept of the celestial sphere from the ancient idea of crystalline spheres. The celestial sphere is an imaginary sphere that is centered on the earth and has dimensions sufficiently large that the earth can be regarded as a point at the sphere's center. Because the astronomer is concerned here only with the *direction* from which the starlight is coming, the images of all celestial objects, regardless of their true distances, are considered to be projected onto the surface of the celestial sphere. Thus, the celestial sphere is very similar to the idealized globe in the latitude-longitude system, and like geographers, astronomers then define a set of reference points and circles on the sphere from which an object's position can be reckoned. Various systems of coordinates have, of course, different sets of reference points and circles.

THE EQUATORIAL SYSTEM. The most commonly used coordinate system in astronomy is the equatorial or *right ascension-declination system* (see Fig. 3.3a).

If we extend the poles of the earth's axis of rotation, the north pole extension will ultimately intersect the celestial sphere at a point that is referred to as the *north celestial pole*. A similar extension of the earth's south pole intersects the celestial sphere at the *south celestial pole*. The *celestial equator* is the great circle on the celestial sphere that is everywhere equidistant from the celestial poles. The *declination* (δ) of an object is defined as the shortest angular distance between the object and the celestial equator. Declination ranges from +90° at the north celestial pole, through 0° at the celestial equator, to −90° at the south celestial pole. *Hour circles* are those great semicircles whose endpoints are the celestial poles.

The prime hour circle chosen as the starting point for the second of the equatorial coordinates is that hour circle which contains the *vernal equinox,* or point where the sun crosses the celestial equator as it passes from south to north in its apparent annual path among the stars (the *ecliptic*). The *right ascension* (α) of an object is then defined as the angle between the prime hour circle and the hour circle of the object. Unlike longitude, the right ascension is always measured from the prime hour circle to the hour circle of the object in a counterclockwise direction looking down from the north celestial pole and ranges from 0^h to 24^h where $1^h = 15°$.

The right ascension and declination of an object can be measured by determining the time and altitude at which the object crosses the celestial meridian, or *transits*. These measurements are accomplished by means of meridian circles which are mounted precisely in the plane of the celestial meridian and which allow the observer access to any por-

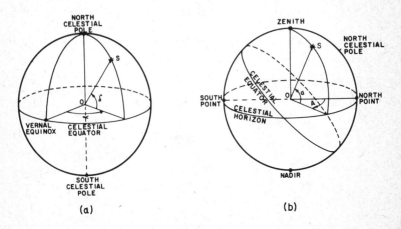

Fig. 3.3 Astronomical coordinate systems: (a) the equatorial or right ascension-declination system and (b) the horizon system. In each case, the position of the star S is specified by two angles.

tion of the celestial meridian lying above the horizon (see below). All the fundamental stellar positions on the celestial sphere are determined by transit observations of one sort or other. Using the fundamental stellar positions as reference points or "celestial surveyor stakes," astronomers can obtain the position of any desired object by photographic means, because the photographic plate provides a reasonable reproduction of the various celestial objects' positions relative to the reference stars.

OTHER NONROTATING COORDINATE SYSTEMS. Although the equatorial system is the most convenient and widely used coordinate system in astronomy, a different reference plane on which to base the system might well have been chosen. Other systems include the *ecliptic system,* which employs the ecliptic as a basic reference plane, and the *galactic system,* which is defined in terms of the mean plane of the Milky Way.

THE HORIZON COORDINATE SYSTEM. In the coordinate systems considered thus far, the points of reference are attached to the celestial sphere, and the coordinates do not change appreciably as a function of the observer's location on the earth or the time of night. Astronomers, however, must measure the coordinates on an earth that to them has a finite size and that is, moreover, rotating with respect to the celestial sphere. To overcome these difficulties, astronomers introduce yet another coordinate system, the *horizon* or *alt-azimuth system* (see Fig. 3.3b). With this system, direct coordinate measurements can be made rather easily, but the values obtained vary significantly with the observer's location on the earth and the time of the observation. This system not only finds a great deal of use in celestial navigation but also serves as a "stepping-stone" coordinate system allowing astronomers to obtain the right ascension and declination of a given object.

The point on the celestial sphere directly above the observer is known as the observer's *zenith.* The point directly below the observer is the observer's *nadir.* The great circle on the celestial sphere that is everywhere equidistant from the zenith and nadir is the *celestial horizon. Vertical circles* are great circles passing through the zenith and nadir. The vertical circle that passes through the north celestial pole is defined as the observer's *celestial meridian.* The celestial meridian intersects the celestial horizon at two points. The point of intersection nearest the north celestial pole is called the *north point* and defines the direction north for the observer.

The coordinates of the horizon system are the altitude and the azimuth. The *altitude* (*a*) of an object is the shortest angular distance between the object and the celestial horizon. The altitude ranges from 0° at the horizon to 90° at the zenith. The *azimuth* (*A*) is the angle between the vertical circle containing the object and the vertical circle containing the north point. It is measured from the north point in a clockwise direction looking down from the zenith and ranges from 0° to 360°.

Celestial Navigation. By the use of astronomical coordinate systems, a navigator can accurately determine his position at sea or in the air from observations of the stars. If the navigator knows the time at a given position on the earth, such as Greenwich, England, and has a table of stellar positions, he can determine his latitude and longitude on the earth's surface by measuring the orientation of these stars relative to his local horizon. For example, if a navigator observed that a star on the celestial equator was at his zenith, he would know that he must be on the earth's equator or at 0° latitude.

Triangulation and Parallax. The accurate measurement of positions also, in some cases, allows astronomers to determine the distance to that object by means of triangulation or parallax as shown in Fig.

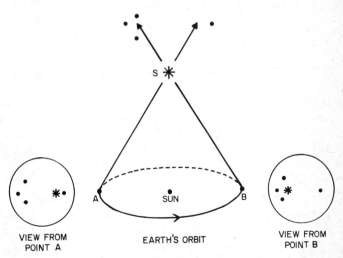

Fig. 3.4 The principle of parallax. As a result of the earth's motion in its orbit, the nearby star S appears to be projected onto a different position of the sky relative to the more distant background stars when viewed from two positions A and B in the earth's orbit.

3.4. By measuring the apparent shift or parallax p of an object viewed from two different locations either on earth or from the earth's orbit, astronomers can obtain trigonometrically the distance r to the object from the relation

$$p \text{ (in arcseconds)} = 206{,}265 \, \frac{D}{r}$$

where D is the distance between the observation points (D must be in the same units as r). If we define the *parsec* (pc) as the distance to an interstellar object that exhibits a parallax of 1 arcsecond when viewed from the earth at right angles to the sun and opposite the sun, then the above relationship becomes

$$p \text{ (in arcseconds)} = \frac{1}{r \text{ (in parsecs)}}$$

Parallaxes provide astronomers with the basic data for stellar distances, since we may also write

$$r = \frac{1}{p}$$

SPECTROSCOPY

When a spectroscope breaks light up into its component wavelengths, three basic types of light patterns or spectra can result: a continuous spectrum, a dark-line or absorption-line spectrum, and a bright-line or emission-line spectrum. Careful analyses of these spectra will yield information to the astronomer concerning the object's composition, temperature, and motion. To understand the principles and techniques involved in the science of analyzing spectra, or spectroscopy, one must first examine the structure of the basic unit of matter in the universe, the atom.

The Structure of the Atom. The smallest unit into which a chemical element can be divided and still maintain the properties of that element is called an *atom*. Each atom consists of a relatively massive nucleus composed of one or more positively charged particles called *protons,* and varying numbers of neutral particles called *neutrons*. Surrounding this heavy nucleus is a system of relatively low-mass, negatively charged particles called *electrons,* which can be thought of as moving in orbits about the nucleus. Each element has a unique *atomic* *number* or number of protons in its nucleus. The element hydrogen, for example, has one proton in its nucleus, helium has two, lithium has three, and so on. Because the masses of the proton and neutron are very nearly the same and the mass of the electron is only about1/1800 that of the proton, it is useful to define the mass of an atom or *atomic mass* in terms of the total number of protons and neutrons present in the nucleus. Thus, a lithium atom with three protons and three neutrons would have an atomic mass of 6. For neutral atoms, the number of electrons and protons is the same, but it is possible for one or more electrons to be stripped from the parent atom to form a positively charged particle called an *ion*.

If the electrons moving about the nucleus behaved as classical particles, they would spin into the nucleus within a very short time. To resolve this difficulty, the Danish physicist Niels Bohr suggested that electrons do not move as classical particles, but are restricted to move in certain orbits or energy levels. The only motion allowed the electron is into and out of these energy levels to other energy levels and, like a ball on a flight of stairs, the electron cannot occupy any position in between the "steps." Moreover, to make these discrete jumps, the electron must either absorb energy to move to a higher level farther from the nucleus or give up energy to move to a lower level closer to the nucleus.

The Continuous Spectrum. In the case of incandescent solids, liquids, and high-density gases, the energy levels of the atoms are crowded to such an extent that their electron energy levels are no longer discretely separated. As a result, it is possible for electrons to make transitions having any amount of energy, and hence the wavelength of the corresponding photon that is emitted can also take on any value. The resulting spectrum of emitted photons contains every possible wavelength and is known as a *continuous spectrum*. This continuum of photon radiation is also referred to as the *electromagnetic spectrum* and is summarized in Fig. 3.5. The human eye detects as a rainbow only that small part of this continuum which lies in the visible region between 4000 Å and 7000 Å. The remaining regions of the electromagnetic spectrum include the gamma rays, X rays, ultraviolet rays, infrared rays, and radio waves and must be detected by less direct means.

Although every energy transition in an incandescent object is possible, not all are equally probable. Thus, the distribution of the total amount of energy emitted at each wavelength is not uniform but will

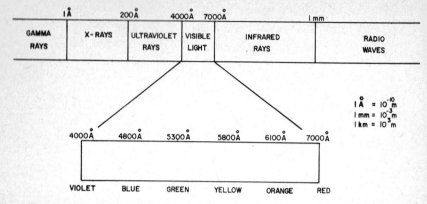

Fig. 3.5 The electromagnetic spectrum.

take on the humped shape shown in Fig. 3.6. Such a graph of energy emitted per unit wavelength (E_λ) plotted against the wavelength (λ) is known as a Planck curve. The actual behavior of this graph can be predicted from the principles of quantum mechanics and is described mathematically by the *Planck function:*

$$E_\lambda = \frac{2hc^2}{\lambda^5}\left(\frac{1}{e^{hc/\lambda kT}-1}\right)$$

in which h is Planck's constant and c is the speed of light waves of any wavelength. The total area under the curve is the total energy output E_{tot} per unit time per unit area emitted by the object at all wavelengths, and is directly proportional to the fourth power of the object's Kelvin temperature, or

$$E_{tot} = \sigma T^4$$

This relation is referred to as the *Stefan-Boltzmann law,* and the constant σ is the Stefan-Boltzmann constant, which in metric units is equal to 5.67×10^{-5} erg/cm^2-deg^4-sec. Thus, if the temperature of an object is doubled, the corresponding increase in its energy output is increased by a factor of 2^4 or 16.

From Fig. 3.6 it can also be seen that the value of E passes through a maximum value. The wavelength λ_{max} at which this occurs is inversely proportional to the Kelvin temperature, or

$$\lambda_{max} = \frac{W}{T}$$

This relation is called *Wein's law,* and the quantity W is the Wein's law constant, which in metric units is equal to 0.29 cm-deg. Astronomers can estimate the temperatures of stars using these laws governing continuous radiation. For example, the wavelength of maximum light for the sun is about 4600 Å or 4.6×10^{-5} cm. From these data, the surface temperature of the sun can be estimated from Wien's law to be $0.29/4.6 \times 10^{-5}$ or about $6300°K$.

Line Spectra. In 1815, the German optician Joseph Fraunhofer noted that the solar spectrum was not a perfectly continuous spectrum but contained several hundred dark gaps or lines. The spectra of the distant stars were found to show a similar effect but with fewer observable lines. It was also discovered that certain types of glowing gases exhibited a bright-line spectrum that consisted of only a few bright lines and no continuum, almost the exact reverse of the dark-line solar spectrum.

These line spectra can be explained qualitatively in terms of the

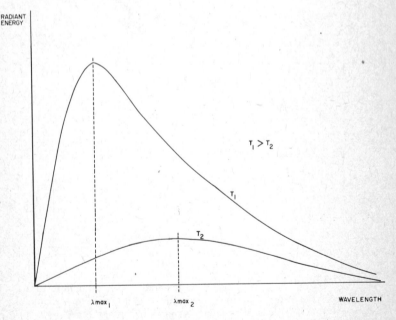

Fig. 3.6 Blackbody or Planck curves for an object at two different temperatures. As the temperature increases, the total energy output (the area under the curve) increases as T^4 and the position of the maximum point decreases as $1/T$.

Bohr model for the atom. If a beam of light containing all possible wavelengths passes through a relatively cool, tenuous layer of gas, the electrons present in the atoms comprising the gas will tap whatever energy from the radiation field they need to move to a higher level in the atom. Since there is a discrete amount of energy involved in such a transition, only those photons having the proper wavelength and corresponding energy will be absorbed by the electrons. As a result, the beam that emerges from the gas will be missing at least some of the

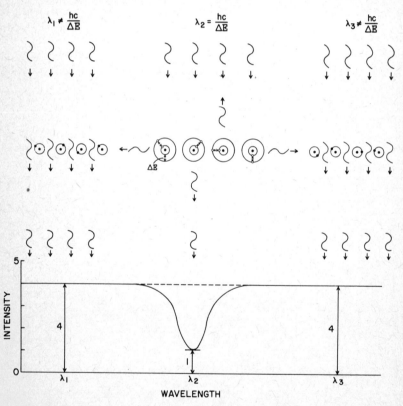

Fig. 3.7 An idealized interaction of a gas and a radiation field to form an absorption line. Photons having wavelengths λ_1 and λ_3 pass through the gas unaffected. The photons at λ_2 have the energy necessary to excite the electrons in the atoms of the gas and are absorbed. Almost immediately the electrons drop back down to their initial levels and reemit the absorbed photons—but in a random direction. Thus there is a reduced intensity at λ_2, which shows up in the object's spectrum as an absorption line.

radiation at a number of discrete wavelengths. If this emergent beam is broken up into a spectrum, the observer will thus see a set of missing wavelengths superimposed on the continuous spectrum that made it through the gas untouched, as shown in Fig. 3.7. This type of spectrum is referred to by astronomers as an *absorption-line, dark-line,* or *Fraunhofer spectrum.*

The electrons that are excited in this fashion, however, generally do not remain in their upper levels or excited states for very long, but drop back down or decay to their initial level. In doing so, the electron must, according to the Bohr model, reemit the energy it absorbed in jumping to its excited state. This energy is usually reemitted in the form of the photon that was initially absorbed. This reemitted photon, however, can go in any direction and is just as likely to go back from whence it came as it is to proceed in its original direction. Thus, a certain number of photons absorbed by the gas are returned to the original beam, but most are reemitted in other directions. An idealized situation for a four-atom system is illustrated in Fig. 3.7. If an observer were stationed in such a way that he could view the continuum source and the cloud separately, as in Fig. 3.8, the light from the cloud would consist only of those photons absorbed by the cloud's atoms and reemitted in the direction of the observer. If this radiation is broken up into a spectrum, no continuum would be observed but rather a set of discrete lines whose wavelengths would correspond to the photon energies absorbed and reemitted by the cloud. Such a spectrum is called

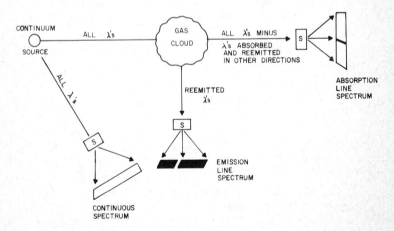

Fig. 3.8 The generation of the three basic types of spectra.

an *emission-line* or *bright-line spectrum*. (The generation of the three types of spectra is diagramed in Fig. 3.8.)

For the astronomer, the importance of line spectra lies in the fact that each substance in the universe has a unique set of energy levels that give rise to a unique set of spectral lines for that substance. Even ions (atoms that have lost or gained one or more electrons) and isotopes (atoms with the same atomic number but different atomic weights) of the same element have slightly different spectral-line patterns. By referring to the spectroscopic "fingerprints" obtained for purified elements here on earth, astronomers can identify which elements are present in the spectra of stars, as well as estimate how much of each element is present.

The task of sorting and analyzing these lines is by no means a simple one. A star, such as the sun, may contain as many as sixty or seventy different elements and molecules, each of which may have several dozen to several hundred spectral lines. Moreover, the strengths of the lines are not determined by element abundance alone, but by several other factors, particularly temperature and degree of ionization, which control how the electrons are distributed within the atom's system of energy levels.

The Doppler Effect. If there is relative motion between a wave source and an observer, the observer will not see the wavelength emitted by the source, but rather a wavelength that is slightly shifted longward or shortward of that originally emitted by the source. This effect, known in physics as the Doppler effect, is illustrated in Fig. 3.9. The component of relative motion that is directed toward or away from the observer is called the radial velocity V_r of the source. The radial velocity for a celestial object can be obtained from the Doppler effect for light waves, using the relationship

$$V_r = \left(\frac{\text{observed } \lambda - \text{laboratory } \lambda}{\text{laboratory } \lambda} \right) c = \frac{\Delta \lambda}{\lambda_{\text{lab}}} c$$

The laboratory wavelengths are essentially the unshifted values emitted from the source and the observed wavelengths are the wavelength values actually measured by the observer. The quantity $\Delta \lambda / \lambda_{\text{lab}}$ is called the Doppler shift and is positive or *red-shifted* if the relative motion is away from the earth and negative or *blue-shifted* if the relative motion is toward the earth.

To determine the direction and magnitude of an object's radial ve-

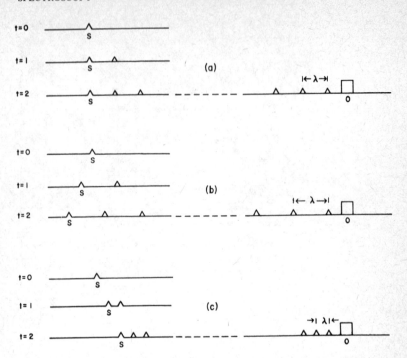

Fig. 3.9 The Doppler effect for (a) no relative motion, (b) relative motion away from the observer, and (c) relative motion toward the observer. Depending on the relative motion between the source S and the observer O, the observed wavelength or distance between successive pulses will either be longer (for relative motion away from) or shorter (for relative motion toward the observer) than that actually emitted by the source.

locity, astronomers superimpose a comparison spectrum, usually an iron arc emission spectrum, onto the same plate as the spectrum of the object, above and below it, as shown in Fig. 3.10. The comparison spectrum is at rest relative to the plate and hence is not Doppler-shifted; the object's spectrum contains the relative motion of the object and its lines will be shifted relative to the comparison spectrum. Because continuum radiation does not readily lend itself to such measurements, astronomers rely solely on absorption and emission lines, whose wavelengths can be measured to ± 0.0001 Å, for the determination of radial velocities.

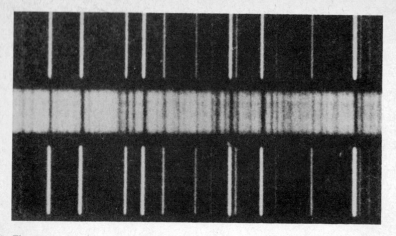

Fig. 3.10 The Doppler shift of spectral lines in the star η Persei. The bright lines above and below the stellar spectrum (center strip) are produced by an iron arc that is at rest relative to the plate and hence serve as unshifted reference points along the spectrogram. (The Kitt Peak National Observatory.)

PHOTOMETRY

The methods and techniques employed by astronomers to determine the brightness of the celestial objects they observe comprise the area of astronomy known as photometry. At first, brightness measurements were made visually by comparing the brightness of a star to some standard star or to a known light source. From the late nineteenth century until after World War II, brightness measurements were made photographically by measuring the size and density or blackening of the image of a given object on a photographic emulsion. Advances in electronics since World War II have led to increased precision in brightness measurement and an almost exclusive reliance by the astronomer on the *photoelectric photometer*.

The Magnitude Scale. Brightnesses of objects in astronomy are measured on a scale of magnitudes that is essentially a rating scale for the energy output of the object being observed. In this scale, called a *Pogson scale*, a difference of five magnitudes corresponds to a ratio of 100:1 in energy. For two objects having arbitrary magnitudes m_1 and m_2, the corresponding ratio of the brightnesses l_1/l_2 is given by

$$m_2 - m_1 = 2.5 \log \left(\frac{l_1}{l_2} \right)$$

A representative sample of magnitude differences and corresponding light ratios is presented in Table 3.1. The magnitude scale is also set up so that the brightest stars in the night sky have an average magnitude of about +1.0 which increases numerically in a positive direction with decreasing brightness. Thus, a second-magnitude star is 100 times brighter than a star of seventh magnitude, and so on. A list of apparent magnitudes for some selected objects is given in Table 3.2. The faintest magnitudes that can be detected by various instruments are presented in Table 3.3.

Photometric Systems. The observed brightness of a given object depends rather intimately on the wavelength sensitivity of the detector. For example, the human eye is most sensitive in the yellow-green region of the spectrum, whereas the typical photographic plate is more sensitive to wavelengths of blue and violet light. Thus, if a blue star and a green star each having the same total energy output are observed

Table 3.1. Light Ratios for Selected Magnitude Differences

$m_2 - m_1$	I_1/I_2
0	1
0.5	1.6
1	2.5
2	6.3
5	100
10	10,000

Table 3.2. Apparent Visual Magnitudes for Some Familiar Objects and Constellations

Object	Apparent Magnitude
Sun	−26.5
Full moon	−12.5
Venus (brightest planet)	− 4.0
Sirius (brightest star)	− 1.5
Big Dipper stars	+ 2.0
Cassiopeia stars	+ 3.0
Pleiades stars	+ 4.0
Uranus	+ 6.0
Neptune	+ 9.0
Pluto	+15.0

Table 3.3.   Limiting Magnitudes for Various Instruments

Instrument	Faintest Detectable Magnitude
Human eye (city)	+ 3.0
Human eye (open country)	+ 6.5
Binoculars	+10.0
10-cm (4-inch) telescope	+12.0
1-meter (40-inch) telescope	+17.0
Largest reflectors	+24.5

with the unaided eye, the green star will appear brighter, whereas on the photographic plate, the blue star will show up as the brighter object. To circumvent this problem, astronomers have developed a number of photometric systems that are defined in terms of the phototube used in the photoelectric photometer and the number and types of filters used on the incoming light beam. The most common of these systems, the so-called U,B,V system, employs three separate filters in coordination with a type 1P21 phototube and allows the astronomer to observe the energy being emitted by a given object in the ultraviolet, blue, and visual or yellow-green regions of the spectrum. Other systems employ different phototubes and different filters to gain sensitivity in other spectral regions.

Color Indexes. Of considerable interest to the astronomer working with a photometric system is the comparative brightness of an object in each of the wavelength regions being observed. This comparison is usually expressed as a magnitude difference called a *color index* or *color,* and in a certain sense multicolor photometry can be thought of as a form of low-dispersion continuum spectroscopy. The most common of the color indices is the B-V color index, which is a comparison of the brightness of an object in the blue region with its brightness in the visual region.

Bolometric Magnitude. The bolometric magnitude of an object is the object's total energy output, at all wavelengths, expressed as a magnitude. Because of the filtering effects of the earth's atmosphere and the limited wavelength sensitivity of the various photometric systems, bolometric magnitudes cannot be determined directly but must be estimated from measurements of the energy output in those wavelength regions of the object's continuum spectrum that are available for observation. Most commonly used in this regard is the *bolometric cor-*

rection, which is the ratio of visual to total energy for an object expressed as a magnitude difference.

Absolute Magnitude. Because the energy per unit area emitted by a given object decreases as the square of the object's distance, the apparent magnitude observed for a given object will depend not only on the object's intrinsic brightness, but also on the distance to the object. To remove the effect of distance on the observed brightness of an object, astronomers define an absolute magnitude M as the magnitude that would be measured if the object were observed at a fixed distance of 10 parsecs. This definition of an absolute magnitude, combined with the inverse square law, yields a relationship between an object's apparent magnitude m, its distance r in parsecs, and its absolute magnitude M, known as the *distance modulus formula:*

$$m - M = 5 \log r - 5$$

If two of the three quantities m, M, and r are known, the other can be determined. Thus, this relationship can be used by astronomers to determine either an object's distance or its intrinsic brightness.

REVIEW QUESTIONS

1. Find the latitude and longtitude of the intersection of the prime meridian and the equator. *Ans.:* $\phi = 0°$, $l = 0°$.
2. What properties do we assign to the celestial sphere? What is the reason for each?
3. Compare and contrast the equatorial and horizon systems of coordinates. What is the use of each?
4. Explain why the altitude and azimuth an observer measures for a given object depend on the time of the observation and on the observer's position on earth, whereas right ascension and declination do not.
5. Summarize the information that can be obtained from the continuous spectrum and the line spectrum of an object.
6. Describe how emission-line and absorption-line spectra are formed.
7. How much brighter is the planet Venus than the planet Uranus? (See Table 3.2.) *Ans.:* 10^4 or 10,000 times.
8. Compare and contrast a photometric color, bolometric magnitude, absolute magnitude, and apparent magnitude.

9. Describe the various methods astronomers can use to determine the brightness of an object.

10. A sodium line is observed in the laboratory to have a wavelength of 5890 Å. If the same line has a measured wavelength of 5893 Å in a certain stellar spectrum, is the star's relative motion toward or away from the earth? What is the relative velocity? *Ans.:* 153 km/sec away from the earth.

4

Motion and Gravity

As we have seen in Chapter 1, the explanation of the motions of celestial objects has occupied a prominent place in the history of astronomical thought. Only since the time of Newton, however, has it been possible to describe these motions with considerable accuracy and detail. This has been accomplished by applying the general laws of motion to the particular problem of movement in gravitational fields.

Before describing motion, it will be useful to review several basic concepts. *Speed* is the amount an object changes its position per unit time. If a direction is assigned to a given speed, the resultant quantity consisting of a specified magnitude and direction is called *velocity*. *Acceleration* is the rate at which a velocity is changing in time. Agents in nature that initiate changes in motion are called *forces*. These agents can arise from a number of sources including electrostatic fields and gravitational fields. The quantity of matter in an object that resists the action of forces is the object's *mass*. Every finite-sized object also possesses a point called the *center of mass,* which behaves as if the entire mass of the object were concentrated at this point. For two bodies mutually revolving about each other the center of mass of the system is often referred to as the *barycenter*.

NEWTON'S LAWS OF MOTION

Almost all phenomena involving motion can be described with three basic statements that Isaac Newton first put forth in his *Principia* in 1687.

Newton's First Law. The first of Newton's laws, the law of inertia, states that any object in a state of rest or uniform motion (constant velocity) tends to resist changes in that state. For example, if an automobile is moving at a high rate of speed and comes to a sudden stop, the passengers' bodies will resist the abrupt decrease in the car's velocity. As a result, they are thrust forward and—without seat belts or harnesses—are subject to serious injury. Because velocity is a vector quantity having both magnitude and direction, the law of inertia also tells us that an object will resist any attempt to change the direction of the motion. Passengers inside a car that turns a corner at constant speed will feel a resistance to the turn.

Newton's Second Law. The law of inertia describes only how an object behaves if no net outside forces are present. Newton's second law describes what happens to an object's motion when an outside force, such as friction, is present. According to this law, the object's response to the outside force will be an acceleration in the direction of the force, the magnitude of which will depend on the object's mass. Simply stated, the second law reads

$$\text{force} = \text{mass} \times \text{acceleration}$$

Note that if the net force is zero, the acceleration is also zero and the resulting motion must be uniform. The law of inertia thus may be regarded as a corollary of the second law. Moreover, if the force law governing a particular effect or phenomenon, such as gravity, can be determined experimentally, then by recalling that accleration and position are related mathematically, it is possible to describe the motion of any particular particle subject to this force law by solving the equation obtained from the second law. The second law also illustrates the difference between mass and weight. *Weight* is the force exerted on a given mass by a gravitational acceleration and changes from place to place in the universe. *Mass,* on the other hand, does not change in a nonrelativistic situation.

Newton's Third Law. The third law of Newton states that if a force (sometimes called an *action*) is exerted on a given object by a second object, the given object will respond with a second force (*reaction*)

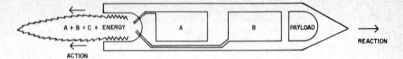

Fig. 4.1 Application of Newton's third law in the design of a rocket. As the chemicals ignite with a rearward release of gases (action), the rocket responds with a forward thrust (reaction).

that is exactly equal and opposite the first. In short, forces always occur in pairs. In some cases, such as the discharge of a shotgun, in which the "kick" is the reaction to the forward thrust of the buckshot, the effect is readily seen. In others, for example, the gravitational force exerted by the earth on a given object, the reactive force (in this instance, the force exerted by the object on the earth) is impossible to observe. One practical application of the third law lies in the design of rocket engines (see Fig. 4.1), in which two chemicals are mixed in an ignition chamber in such a way that the force of the explosion is directed out an aperture at the rear of the rocket. The rocket responds to this force by moving forward.

CELESTIAL MECHANICS

Celestial mechanics is that branch of astronomy which deals with the motions of the members of the solar system.

The Law of Universal Gravitation. In addition to his laws of motion, Newton's *Principia* contains this very important discovery:

Between any two objects anywhere in space, there exists a force of attraction that is in proportion to the product of the masses of the objects and in inverse proportion to the square of the distance between them.

This statement is known as the law of universal gravitation and can be written mathematically as

$$F_g = G \frac{\mathbf{m}_1 \mathbf{m}_2}{r^2}$$

where $\mathbf{m}_1$ and $\mathbf{m}_2$ are the two attracting masses*, r is the distance between them, and F_g is the resultant gravitational force between them.

*The symbol $\mathbf{m}$ will be used throughout the text to denote mass and to distinguish it from the symbol for magnitude.

The constant of proportionality G is the universal gravitational constant.

The law of universal gravitation is an experimentally determined or empirical relationship and cannot be derived from any theory now available. It can, however, be used in conjunction with Newton's second law to examine the behavior of bodies moving under the influence of gravitational fields. Such a study, when applied to astronomical objects, is referred to as celestial mechanics.

Kepler's Laws. If the laws of motion are developed mathematically from Newton's second law and the law of gravitation for two objects moving in each other's gravitational fields, three basic statements can be made regarding the behavior of the system. These statements are more generalized versions of the laws Kepler arrived at empirically early in the seventeenth century (see Chapter 2).

Law 1. The orbit or path of one object about the other will take one of the so-called conic section curves, which means that the path must be a circle, ellipse, parabola, or hyperbola (see Fig. 4.2).

Law 2 (Law of Areas). One object will move about the other in such a way that the line joining the two objects will sweep out equal areas of the orbit over equal time intervals (see Fig. 4.3).

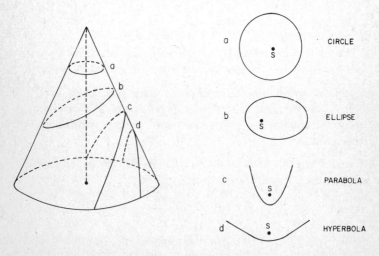

Fig. 4.2 The conic sections. Kepler's first law requires that any orbit resulting from a two-body gravitational interaction must take on one of these shapes.

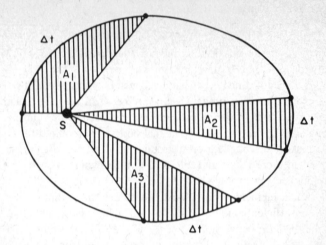

Fig. 4.3 The law of equal areas. If the time interval Δt is the same, the corresponding areas swept out by the planet in any other Δt are the same.

Law 3 (Harmonic Law). The time taken by one object to orbit the other (sidereal period, P) is related to the size of the orbit (mean distance, a) by the following:

$$\frac{a^3}{P^2} = \frac{G}{4\pi^2}(\mathbf{m}_1 + \mathbf{m}_2)$$

It is laws 1 and 3 that are somewhat different from those originally formulated by Kepler. Since all the planets in the solar system are moving in elliptical orbits about the sun, Kepler's original formulations of laws 1 and 3 do not allow for any nonelliptical paths. Kepler's law 3 did not take into account the masses of the planets; however, the sun is so massive compared to the planets that any planet's contribution to the total mass of a sun-planet two-body system can be neglected. As a result, the ratio a^3/P^2 is very nearly equal to the product of $G/4\pi^2$ and the mass of the sun for *all* the planets. For convenience, in calculations involving law 3, the mean distance a is expressed in units of the mean earth-sun distance called the *astronomical unit* (see p. 59), P is expressed in *earth years,* and $\mathbf{m}$ is expressed in *solar masses* (the sun's mass). For the earth-sun system, the value of $G/4\pi^2$ using these units is thus unity, and since $G/4\pi^2$ is a universal constant, it will remain equal to one for any other system as long as the above system of units is preserved.

The Elements of Orbits. To predict an object's position in its orbital path, the orientation, size, and shape of the orbit in space must be specified. For an object orbiting the sun, the specifications, known as orbital elements, include three orientation angles, i, Ω, and ω, for the orbit (see Fig. 4.4); the eccentricity e, which is a measure of the shape of the orbit; the mean orbital distance a between the object and the sun; the exact time of the object's closest approach to the sun (perihelion passage, I_o); and the time (sidereal period, P) required for an object to make one complete orbit around the sun. These elements are described in more detail in Table 4.1. Orbital elements such as these can be defined for any two-body system.

The orbital elements of an object can be determined in a variety of ways. If a sufficient amount of time is available, it is possible to determine the distance between the object and the sun for a number of the object's orbital positions, using the procedures outlined in Fig. 4.5 to obtain each position. Historically, these were the methods employed by Kepler to obtain his planetary orbits using the data assembled by Tycho Brahe. To determine an orbit using data obtained over a more restricted time period requires an advanced knowledge of celestial

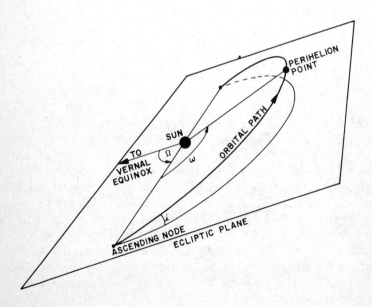

Fig. 4.4 The orbital elements i, ω, and Ω.

Table 4.1. The Elements of an Orbit About the Sun

Orbital Element	Symbol	Description
Orbital inclination	i	Angle between the plane of the object's orbit and the plane of the earth's orbit (see Fig. 4.4).
Longitude of the ascending node	Ω	Angle between the vernal equinox point and the point (ascending node) where the object crosses the plane of of the earth's orbit from south to north. Measured along the earth's orbital plane (ecliptic) from the vernal equinox point and in a counterclockwise direction looking down from the north (see Fig. 4.4).
Argument of perihelion	ω	Angle between the perihelion point (nearest point to the sun) of the orbit and the ascending node. Measured along the plane of the object's orbit from the ascending node and in the direction of the object's motion (see Fig. 4.4).
Eccentricity	e	Dimensionless measure of the shape of the orbit, given by the relation $$e = 1 - \frac{\text{perihelion distance}}{\text{mean distance}}$$
Mean distance, or semimajor axis	a	Average distance between the object and the sun over one complete orbit.
Time of perihelion passage	T_0	Exact time at which the object is closest to the sun (perihelion). One such perihelion passage occurs for each complete orbit.
Sidereal period	P	The time required for the object to make one complete orbit of the sun.

mechanics, but in principle it can be obtained by making three position measurements of an object at three different times.

The Astronomical Unit. It is relatively easy by trigonometry to determine planetary diameters and the scale of planetary distances in the solar system in terms of the astronomical unit (AU). Unfortunately, the astronomical unit as a solar system yardstick is of limited value unless it can be expressed in terms of an earth-defined unit of length

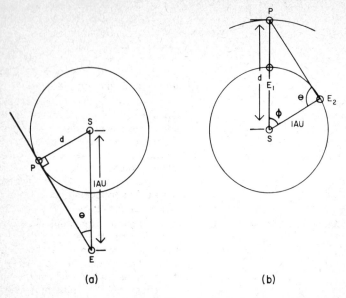

(a)   (b)

Fig. 4.5 Determining planetary distances in AU for (a) an interior planet at greatest elongation and (b) an exterior planet. In Fig. 4.5b, the exterior planet is observed at opposition when the earth is at E_1 and again one sidereal period later when the planet has returned to P and the earth is now at E_2. Both triangles in (a) and (b) are easily solved trigonometrically for the distance d.

such as the mile or kilometer. To find the length of the astronomical unit, astronomers have developed a number of methods by which the distance to an object in the solar system can be measured simultaneously in kilometers and astronomical units.

Occasionally an asteroid will pass sufficiently close to the earth that it will exhibit a parallax, or apparent displacement, to a pair of observers widely separated on the earth's surface. The earth-asteroid distance in kilometers can be determined if the separation of the observers is known and the parallactic shift is measured accurately. The distance in astronomical units can be determined independently from celestial mechanics for the time of the observation, and the length of the astronomical unit in kilometers is thus obtained. Until recently, the most accurate measurement of the length of the astronomical unit was made using this method on the asteroid Eros when it passed within 26 million km of the earth in 1931.

The best value currently available for the length of the astronomical unit, however, was obtained by sending radar waves to the distant planets. The time from the initial pulse t_0 to the observed return echo t_e represents that total time for the radar wave to travel to and from the planet, and because radar waves are known to move at the speed of light c, the distance to the planet is $[(t_e - t_0)/2] c$. Because the velocity of light is one of the most accurately known constants in nature, this method has allowed astronomers to determine the length of the astronomical unit to within 3 km out of a value of some 149,597,870 km.

Once the length of the astronomical unit is known accurately, the mass of the sun and hence planetary masses can be determined in terms of earth-defined mass units, such as the gram, by applying Kepler's harmonic law to the earth-sun system:

$$\frac{a^3}{P^2} = \frac{G}{4\pi^2} (\mathbf{m}_1 + \mathbf{m}_2)$$

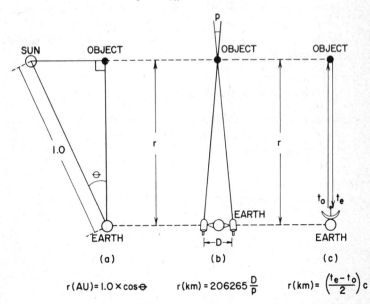

Fig. 4.6 To determine the length of the astronomical unit (AU), the distance to an object must be determined simultaneously in AU and in kilometers. The former can be obtained by simple trigonometric means (a), whereas the latter is usually accomplished by (b) geocentric parallax measurements or (c) radar ranging measurements.

where a, P, and G are all measurable in earth-defined units. Using this method the mass of the sun is found to be about 2×10^{33} grams. Any planetary mass determined in solar masses can thus be converted to grams through the use of this value.

Artificial Satellites. Although the first artificial satellite was not launched until 1957, Newton had speculated on the possibility more than two centuries earlier. From the laws of motion it can be demonstrated that the rate at which an object falls to earth does not depend on the object's horizontal velocity. Thus, a cannonball fired in a horizontal direction will fall to the ground in exactly the same time as if it were dropped from the same height with no horizontal velocity, but it will, of course, go farther. If an object is fired at a high enough, or critical, horizontal speed, it will continuously fall back toward the earth but will never be able to reach the earth's surface because of the combined effect of the earth's curvature and the object's horizontal velocity (see Fig. 4.7). The object then becomes an artificial satellite that will orbit the earth as long as its velocity is not slowed to the extent that its orbital path will intersect the earth's surface. Such an earth orbit is achieved when the horizontal velocity of a projectile exceeds about 28,800 km/hr.

APPARENT MOTION IN THE SOLAR SYSTEM

Because they are viewed from an off-center position, the sun, moon, and planets exhibit various patterns of apparent motion to the earth-based astronomer.

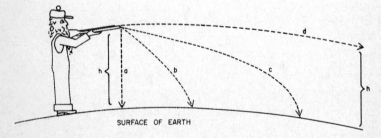

Fig. 4.7 The mechanics of an artificial satellite. If the farmer fires a shell progressively faster (paths a, b, and c), the shell will hit the ground ever farther from the farmer. The shell in path d moves so fast in a horizontal direction that the earth's surface curves away from its path of fall. The bullet thus continuously "falls" around the earth in an orbital path.

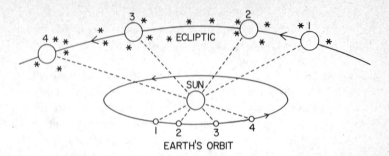

Fig. 4.8 The generation of the ecliptic. As the earth moves around the sun, the sun's image appears to be projected on various star groups. The path of the sun's center so generated is defined as the ecliptic.

The Apparent Path of the Sun. As the earth moves about the sun, the solar glare blots out various background star groups, thus rendering the constellations in that area of the sky invisible. Hence, Orion is visible in the winter sky because the sun appears in the part of the sky that is opposite Orion during the winter months. Six months later in the summer, however, the sun is projected onto the constellation Gemini, which is just to the north of Orion, and Orion disappears into the solar glare. This apparent annual path along which the sun moves among the stars is called the *ecliptic* (see Fig. 4.8), and the constellations through which the sun passes on its yearly journey around the heavens are the *signs of the zodiac*. Traditionally, the signs of the zodiac are the following constellations:

Pisces	Gemini	Virgo	Sagittarius
Aries	Cancer	Libra	Capricornus
Taurus	Leo	Scorpio	Aquarius

Technically, using the 1928 IAU constellation boundaries, the ecliptic also passes through the constellations of Ophiuchus and Cetus.

The time required for the sun to complete one such revolution of the celestial sphere along the ecliptic is thus exactly the same as the time it takes the earth to complete one lap in its orbit. This interval defines another important unit of time, the year, which contains slightly less than 365¼ days.

Apparent Paths of the Planets. The apparent motions of the planets as seen by an earth-based observer are somewhat different

from those described above for the sun. Because of the planar symmetry of the solar system, the apparent motions of all of the planets
are confined to a relatively narrow zone of the sky centered on the
ecliptic, known as the *zodiac*.

INTERIOR PLANETS. Two of the planets, Mercury and Venus, have
orbits that lie entirely inside that of the earth. These *interior planets*
are thus acted on more strongly by the sun's gravity and also have a
shorter distance to travel in order to make a complete circuit of the
sun. As a result, these planets catch up to and pass the earth at regular
intervals, and as seen from the earth they appear to move alternately
from one side of the sun to the other. When Mercury or Venus reaches
its greatest angular separation from the sun, to the east or to the west,
it is said to be at *greatest elongation*. At *greatest western elongation*
the planet appears in the predawn morning sky as a morning star. At
greatest eastern elongation, the planet lies to the east of the sun, sets
after the sun, and is thus visible in the evening twilight as an evening
star. When an interior planet passes between the earth and the sun, it
is said to be in *inferior conjunction;* it is in *superior conjunction* when
it is aligned with the earth and sun on the far side of the sun (see Fig.
4.9).

EXTERIOR PLANETS. The remaining planets' orbital paths lie entirely
outside of the earth's orbit, and these planets, called the *exterior
planets,* exhibit a much different type of motion as seen from the
earth. Most of the time these planets move in a west-to-east direction
across the sky just as the sun and moon do. Every so often, however,
an exterior planet will stop its easterly motion, move backward for
several weeks, stop again, and then resume its normal motion. Such a
phenomenon, called *retrograde motion,* is displayed by every exterior
planet in varying degrees at various times. The explanation of this

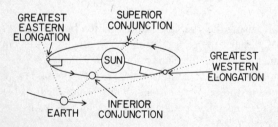

Fig. 4.9 Configurations for an interior planet.

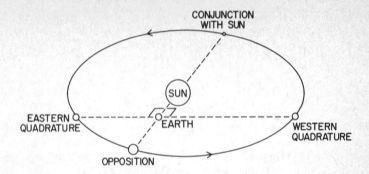

Fig. 4.10 Configurations for an exterior planet.

movement was a topic of considerable debate throughout much of astronomical history. We now know it is caused by the earth's catching up to and passing these planets in their journeys about the sun; when this occurs it is the earth that is more strongly affected by the sun's gravity and has the shorter orbital circumference to travel.

Like the interior planets, the exterior planets have a set of significant configurations with the sun as viewed from the earth. The most important of these occurs when the planet is exactly opposite the sun in the night sky. At this point, called *opposition,* the planet is closest to the earth and is thus most suitably situated for astronomical observations (see Fig. 4.10). If the planet is at right angles to the sun, it is said to be at *quadrature.* Like greatest elongation for the interior planets, quadrature can occur to the west of the sun (western quadrature) or to the east of the sun (eastern quadrature). When an exterior planet lies on the far side of the sun as seen from the earth, it is said to be in conjunction with the sun, a *conjunction* being any close approach of one celestial object to another.

SYNODIC AND SIDEREAL PERIODS. The time necessary for a planet to return to a particular configuration is called its *synodic period* and is important if one wishes to know when and where to look for a given planet in the night sky. For example, the synodic period of the planet Mercury is about 116 days. Thus, if Mercury is observed as an evening star at a particular time, one would again look for the planet in the evening sky some four months later. The synodic period is also important to the astronomer because it can be used to calculate the *sidereal period* using the following equations:

for an interior planet,

$$\frac{1}{\text{planet's sidereal period}} = \frac{1}{\text{earth's sidereal period}} + \frac{1}{\text{planet's synodic period}}$$

and for an exterior planet,

$$\frac{1}{\text{planet's sidereal period}} = \frac{1}{\text{earth's sidereal period}} - \frac{1}{\text{planet's synodic period}}$$

The sidereal and synodic periods are usually measured in days or years, but all periods must be expressed in the same units.

GRAVITY WAVES

For a number of years, theoretical physicists have predicted the existence of gravity radiation or gravity waves that would arise from a rapid periodic motion of masses such as might occur in a close binary star system. Gravity waves would be analogous to the electromagnetic waves generated by oscillating electrons. Calculations by astronomers and physicists indicated, however, that such waves would be so weak (approximately 10^{36} times weaker than electromagnetic waves!) that it would be practically impossible to detect their presence. In 1969, however, a University of Maryland team of physicists headed by Dr. Joseph Weber designed a set of instruments with which, after over a decade of research, they were able to detect the elusive gravity waves. Attempts by other research groups, however, have failed to reproduce Weber's results.

REVIEW QUESTIONS

1. Compare and contrast the following: acceleration, velocity, speed.
2. State Newton's laws of motion and give an everyday example of each.
3. What is the law of gravity?
4. Describe Kepler's laws of planetary motion. How are they related to Newton's laws?
5. How is an orbit described in space?
6. The mean distance between Venus and the sun is about 0.72 AU. If a radar echo is observed to return from Venus 280 seconds after

the initial pulse when Venus is at inferior conjunction, find the length of the astronomical unit. *Ans.:* 1.5×10^8 km.

7. Describe how an artificial satellite is launched into orbit.

8. The moon circles the earth at a distance of about 0.003 AU once every 0.08 year. Find the mass of the earth-moon system in solar masses. *Ans.:* 4×10^{-6} solar mass.

9. What is the difference between the ecliptic and the zodiac?

10. Why can an interior planet never be observed at quadrature or opposition?

11. Calculate the sidereal period of Mars if the interval between successive oppositions is 2.14 years. *Ans.:* 1.88 year.

12. What is the difference beween a synodic and a sidereal period of a planet?

13. What are gravity waves?

5

The System of Planets

The triumph of the heliocentric system in the seventeenth century firmly established the earth as one of six planets swinging about the sun in vast orbits millions of kilometers in diameter. Since that time, three additional planets have been discovered by telescopic means, bringing the total of known planets orbiting the sun to nine.

MERCURY

Most elusive of the naked-eye planets is Mercury. Because of its large orbital speed and close proximity to the sun, Mercury is seldom observed in total darkness. Despite these difficulties, however, astronomers have managed to assemble a fairly impressive array of data concerning this tiny world.

Motion. Mercury moves in a slightly flattened elliptical orbit about the sun at a mean distance of 58 million km once every 88 days, thus making it the closest planet to the sun. Because its orbit lies entirely within that of the earth, Mercury exhibits a complete set of phases similar to those observed for the earth's moon. Mercury shows phases, however, because we view its sunlit side from various angles as it circles the sun. Venus displays the same effect, and this observation prompted Galileo to assert correctly that both planets circled the sun.

On occasion, Mercury passes directly between the earth and the sun. Such an event, called a *transit,* occurs for Mercury in intervals of 3, 7, 10, or 13 years. The last such transit occurred on November 10, 1973, and the next is due on November 13, 1986.

Mercury's orbital motion is of considerable interest to astronomers because the line joining the near and far points of the orbit, the *line of apsides,* slowly shifts its orientation in space, or precesses. Approximately 43 arcseconds per century of this motion cannot be accounted for in terms of classical mechanics. One attempt to do so led to the ''discovery'' and subsequent rejection of a planet prematurely named Vulcan lying between the orbit of Mercury and the sun. It is now known that the precession in Mercury's orbital orientation is due to a small effect in the planet's motion that can be predicted and calculated from relativity theory. There have been recent suggestions that some of the precessional motion is due to a slight flattening of the solar disk, in which case certain aspects of relativity theory would have to be altered. Because of the extreme difficulty in measuring some of the effects involved, this issue is still in doubt.

A second aspect of Mercury's motion, its rotation, has also been the subject of debate. Visual observations of the planet's somewhat indistinct markings had for many years led to the conclusion that Mercury's rotational period was the same as its orbital period and that the planet always kept the same hemisphere facing the sun, much as the moon faces the earth. Radio and radar observations made in the late 1950s and throughout the 1960s revealed a somewhat different picture of Mercury's rotation. The dark-side temperature of Mercury obtained at radio wavelengths was much higher than would be expected if the dark side were perpetually facing away from the sun. Moreover, radar beams reflected from the planet's edges exhibited a Doppler shift that could be accounted for only if the planet were spinning on its axis once every 59 days or so. Subsequent investigations have also shown that some of the features used visually to determine Mercury's rotation rate may have been misidentified, and as a result, the 59-day rotation period is the currently accepted value.

Satellites. Mercury has no known natural satellites.

Physical Properties. Mercury is the smallest of the known planets and has a measured diameter of some 4840 km. Because it has no satellites, its mass must be determined from its gravitational effects on closely passing objects such as comets and asteroids. Such studies yield a mass for Mercury of about one-twentieth of an earth mass. The

Fig. 5.1 Space probe
view of Mercury. (NASA.)

mean density or total mass/total volume of Mercury is then found to be
5.4 g/cm^3. The reflecting power or *albedo* of Mercury is very low
(about 0.06) and suggests that Mercury possesses an airless, rugged
surface like that of the moon.

Being so close to the sun, Mercury has a rather high surface temperature. It also has a weak gravitational field. These facts lead astronomers to believe that whatever atmosphere Mercury may once
have had has long since boiled off into space. No definitive evidence
to the contrary has been presented to date.

Visual observations of the planet's surface show hints of dusky features similiar to the lunar maria (see Chapter 7), and observations of
reflected radar signals from Mercury can best be accounted for by assuming that the planet's surface is very rough. This latter observation
has been confirmed by Mariner space-probe photographs of the
planet's surface (see Fig. 5.1). Additional data from the Mariner space
probes indicate that Mercury possesses a weak magnetic field, about
$1/100$ as strong as that of the earth. There is also considerable evidence
that Mercury contains a rather large iron core.

Fig. 5.2 Space probe view of Venus. (NASA.)

VENUS

Aside from the sun and the moon, the most beautiful and impressive celestial object that can be seen with the naked eye is Venus. However, the planet lies largely in mystery, for nearly all of its properties must be unlocked from the dense cloud layers that surround its surface.

Motion. Venus moves in an almost perfectly circular orbit once every 225 days at a distance of 108 million km from the sun. Venus thus can come closer to the earth (41 million km) than any other planet. Venus's orbit, like Mercury's, lies entirely within that of the earth, and hence not only does Venus display a complete set of phases which are readily observable in even a small-aperture telescope, but it also is able to transit the sun's disk about every 120 years. Transits of Venus usually occur in pairs; the last such pair occurred in 1874 and 1882, and the next pair is due in 2004 and 2012

Venus's surface is shrouded in clouds, thus precluding direct measurement of its period of rotation (see Fig. 5.2). Radar measurements of the Doppler effect in the radar signals reflected from the edges of the planet indicate that Venus spins on its axis once every 243 days in a retrograde or "backward" sense, that is, east to west.

Satellites. Venus, like Mercury, has no known natural satellites.

Physical Properties. Venus is some 12,200 km in diameter or only slightly smaller than the earth. Although Venus has no known natural satellites, its mass, like that of Mercury, can be determined from its grativational effects on celestial objects that pass relatively close. Particularly good values for the mass have been determined from the effects of Venus's gravity on the Mariner fly-bys and the Soviet Venera soft-landers. The mass so obtained is 0.82 earth mass. The mean density of Venus is thus some 5.3 g/cm^3.

When viewed through a telescope, Venus displays no permanent markings, but only a series of dusky patches that change rapidly from day to day in both size and shape. This fact, combined with a very large observed albedo (0.76), suggests that Venus possesses a dense, opaque atmosphere. Spectroscopic observations show that Venus's atmosphere is about 90 percent carbon dioxide, 7 percent nitrogen, 1 percent oxygen, and traces of other gases, including water vapor and nitrogen oxides. Data from the Soviet Venera soft-landers indicate a surface pressure of 10 to 20 atmospheres, a pressure equivalent to that found several hundred meters beneath the ocean, and a temperature of 600° to 700°K, or nearly that measured for the surface of Mercury. It is believed that this high degree of heating is the result of the trapping of solar radiation in Venus's atmosphere by means of the so-called greenhouse effect (see Fig. 5.3). Despite the opacity of the atmosphere in the visible and infrared regions of the spectrum, longer-wavelength radio and radar waves can pass relatively unimpeded through the cloud layers. The reflection of longer-wavelength radiation sent to Venus strongly suggest that the planet's surface is very rough. This view has been confirmed to some extent by the photographs of the Venusian surface sent back to earth by the Soviet soft-landers Venera 9 and 10.

THE EARTH

Because, as the home planet, the earth has been studied so extensively, it will be discussed in a separate chapter. However, a brief description of the earth-moon system will be presented here for purposes of comparison with other worlds in the solar system.

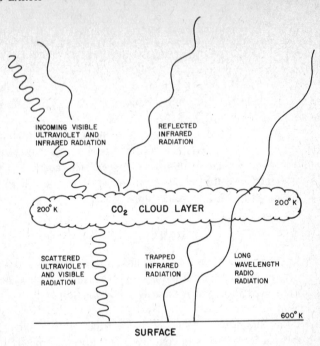

Fig. 5.3 The greenhouse effect in the atmosphere of Venus. The radiation trapped beneath the carbon dioxide cloud layer serves to heat the surface to an extremely high temperature.

Motion. The earth moves in a slightly elliptical orbit at a mean distance of 150 million km from the sun. The time required for the earth to complete a revolution of the sun is defined as one *sidereal* year. The earth also spins on its axis, and the time it takes for one complete rotation is defined as the day.

Satellites. The earth has only one satellite, the moon, which is so large with respect to the earth that the earth-moon system is often regarded as a double planet.

Physical Properties. Satellite observations indicate that the earth is essentially an oblate spheroid 12,714 km in diameter at the poles and 12,756 km at the equator, and has a mass of about 6×10^{27} grams or about 3×10^{-6} solar mass.

The albedo of the earth is about 0.35. To an extraterrestrial observer, this would suggest the presence of a significant atmosphere, which would be confirmed by observations of an extensive cloud cover

over much of the earth's surface. The earth's atmosphere is composed of nitrogen (78 percent), oxygen (21 percent), argon (0.9 percent), and trace amounts of water vapor, carbon dioxide, and a number of other gaseous substances.

A unique feature of the earth is the fact that nearly three-fourths of its surface is covered by water ranging in depth from a few meters to over 10 km.

The solid surface of the earth consists of high, rugged mountain ranges, flat plains, great deserts, and a pair of polar ice caps thousands of feet thick. Moreover, the earth's surface is still very active geologically, as evidenced by the relatively frequent occurrence of earthquakes, volcanic eruptions, and other upheavals.

MARS

No object in the night sky has inspired more interest and controversy than the planet Mars. It is the one object in the solar system, aside from the earth, for which there is a reasonable possibility that life forms may be present.

Motion. Like the orbit of Mercury, the Martian orbit is somewhat eccentric and carries Mars from roughly 200 million km out to 250 million km from the sun over a period of some 687 days. At times of opposition the distance between Mars and the earth can thus range anywhere between 50 and 100 million km, and the oppositions of especially close approaches to the earth, which occur at 15- to 17-year intervals, are called favorable oppositions. The last favorable opposition of Mars occurred in 1971, and the next is due in 1988.

From direct observations of features on Mars's visible surface, the planet's rotation period is found to be 24^h37^m, or slightly longer than that of the earth. Like the earth, Mars has an axial tilt of about 24°, which gives rise to a similar set of seasons.

Satellites. Mars has two natural satellites, Phobos and Deimos, which were discovered in 1877 by Asaph Hall. They circle their primary at 9300 km and 23,400 km from the center of Mars with respective sidereal periods of 7^h39^m and 30^h18^m. Photographic data obtained by the Mariner space probes show both satellites to be dark, cratered chunks of rocky material which are somewhat oblong in shape (see Fig. 5.4). Phobos is about 20×28 km in size, and Deimos, the smallest known satellite in the solar system, has dimension of only 10×16 km.

Fig. 5.4 Mars's satellites, Phobos and Deimos, are visible from earth only as starry specks close to Mars (above). Space probe close-ups (below) show Phobos (left) and Deimos (right) to be cratered chunks of debris. (Above: Courtesy McDonald Observatory of the University of Texas; below: NASA.)

Physical Properties. Mars's diameter is 6760 km. From the motions of its satellites and from its gravitational effects on space probes, its mass is calculated to be one-tenth that of the earth. The mean density of Mars is 3.9 g/cm^3, and the mean surface temperature is some 250°K.

Mars is the only planet in the solar system whose surface is readily accessible to earth-based astronomers. Telescopically, Mars displays a reddish disk interlaced with irregularly shaped dark areas. Also observed are white polar caps, each of which increases in size in the

Martian autumn and winter for the given hemisphere and recedes in the spring and summer.

Closeup shots of the Martian surface (Fig. 5.5) taken by the Mariner space probes reveal a wealth of geological features. Most spectacular, perhaps, is an enormous "Grand Canyon" rift valley dozens of kilometers wide and over 7000 meters deep which stretches more than 3200 km across the Martian equatorial regions. Also to be seen are volcanic caldera as large or larger than any found on the earth, numerous impact craters, and long sinuous valleys resembling the lunar rilles. The rille-like features are especially interesting because, unlike those found on the moon, they could well have been created by flowing water relatively recently in Mars's geological history. The Martian polar caps are almost certainly a combination of ice and frozen carbon

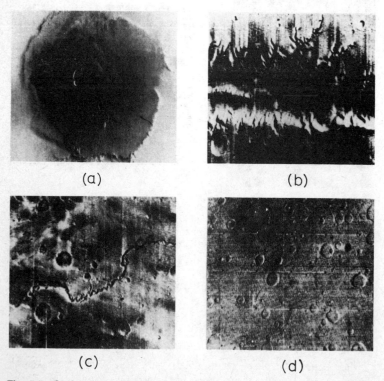

(a)   (b)

(c)   (d)

Fig. 5.5 Surface features on Mars: (a) a volcanic caldera, (b) the gigantic chasm in Tithonicus Lacus, (c) a Martian "rill," and (d) a field of impact craters. (NASA.)

dioxide, or dry ice; they are roughly 1 km thick and about 640 km in diameter during the Martian winter. The famous "canals" of Mars do not appear on the Mariner shots and are now generally held to be optical illusions resulting from the inability of earth-based telescopes to resolve greater detail on the planet's surface.

Mars's low albedo (about 0.15) indicates that at best a thin atmosphere surrounds the planet. More detailed observations of Mars show that its atmosphere is about $^1/_{100}$ as dense as that of the earth. The Martian atmosphere is also known to contain mostly carbon dioxide and some water vapor. Trace amounts of nitrogen, argon, and oxygen, have also been reported by the Viking I and II soft-landers. The Martian atmosphere is capable of raising gigantic windstorms with velocities of more than 480 km/hr that extend over almost the entire surface for weeks at a time. It is believed that such windstorms have been the most important agent in carving and eroding the present surface of the planet.

Viking analyses of the Martian soil have shown it to be rich in iron (13–15 percent), silicon (14–24 percent), calcium (3–5 percent), aluminum (2–5 percent), and titanium ($^1/_3$–$^1/_2$ percent). Tests designed to detect life on Mars indicate that the Martian soil is chemically active. Whether this activity is due to the presence of life forms or to some other type of chemical response has not yet been resolved.

JUPITER

In contrast to the compact, low-mass, high-density "terrestrial" planets out to and including the planet Mars, the worlds beyond Mars are characterized by relatively large masses, large diameters, and very low densities. Typical of these "gas giants" is the largest planet in the solar system, Jupiter (see Fig. 5.6).

Motion. Jupiter moves about the sun once every 12 years in a nearly circular orbit with a radius of 5.2 AU or about 778 million km.

No permanent surface features can be detected on Jupiter, owing to its deep, opaque atmosphere. Observations of the Doppler effect in the light reflected from the extremities of the planet indicate that, on the average, the planet completes a single rotation once every 9^h55^m, but does not rotate as a rigid body. Because of this extremely rapid rate of rotation, the planet is flattened to about one part in fifteen. The earth, by contrast, is flattened to about one part in three hundred.

Satellites. Because of its enormous mass, Jupiter has assembled an

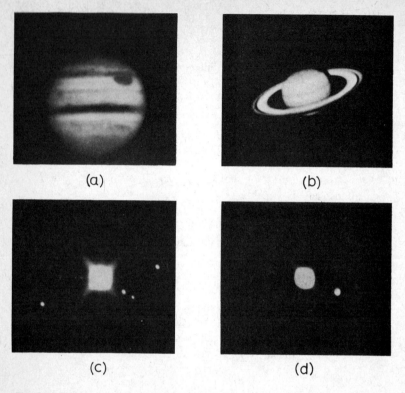

Fig. 5.6 The gas giants: (a) Jupiter, (b) Saturn, (c) Uranus, (d) Neptune. The smaller images in (c) and (d) are satellites. [(a) Courtesy of the Hale Observatories; (b) Lick Observatory photographs; (c) and (d) Yerkes Observatory photographs.]

impressive entourage of fourteen known satellites, including the four so-called Galilean satellites that were first discovered by Galileo in 1610 and were used at that time to provide strong indirect support for the heliocentric theory. Two of these satellites, Ganymede and Callisto, are actually as large as or larger than the planet Mercury, and Ganymede may even possess a thin atmosphere. Five of the outermost satellites have orbits that are highly inclined to Jupiter's orbital plane and are moving in a retrograde direction. It is generally believed that these satellites are asteroids (see Chapter 7) that have been gravitationally captured by the massive Jupiter.

Physical Properties. From the motions of Jupiter's satellites and the planet's gravitational effect on other planets in the solar system, astronomers have found that Jupiter is 318 times more massive than the earth, a mass that is greater than that of all the other planets combined. Jupiter also possesses the greatest diameter (142,700 km) of any planet in the solar system. Despite its large mass, however, Jupiter's mean density is only about 1.3 times that of water, a fact which suggests that the planet is composed primarily of lighter materials, especially hydrogen and helium. It is believed that the pressures that exist in the deeper layers of Jupiter are sufficient to compress both hydrogen and helium into their respective solid states. Thus, unlike the sun, which is known to be gaseous throughout, Jupiter may have a solid core. One interesting observation in this regard is the discovery that despite a low (130° K) outer temperature and an albedo of 0.5, Jupiter actually radiates twice as much energy into space as it receives from the sun, prompting some astronomers to speculate that Jupiter may be the second component of a binary star system with the sun that was not massive enough to ignite into a full-fleged star.

Most intriguing of all, however, is Jupiter's atmosphere which, from spectroscopic observations, is known to consist primarily of hydrogen (70 percent), with methane (0.3 percent) and ammonia (0.1 percent) present in smaller amounts. Measurements of the dimming of stars occulted by Jupiter's atmosphere indicate that helium is also a major constituent of the Jovian atmosphere (20–30 percent).

Jupiter's atmosphere is known to be in an exceedingly active state. High-velocity atmospheric currents similiar to the jet streams on the earth superimposed on a welling up and sinking circulation are believed to be responsible for the series of ever-changing bands that are observed to run parallel to the planet's equator. In 1955, the planet was found to emit radio radiation which, unlike radio radiation arising from a normal blackbody, increased in intensity with increasing wavelengths. Much of this nonthermal radiation can be accounted for by charged particles from the interplanetary medium being trapped and accelerated by a Jovian magnetic field over 100 times as strong as our own terrestrial field. The situation is believed to be an exaggerated version of the earth's Van Allen belts (see Chapter 6). Many aspects of Jupiter's radio emissions, however, cannot be accounted for at present.

Features believed to be atmospheric disturbances are also observed to appear suddenly on the planet's disk and dissipate over several weeks or months. The most famous of these is the so-called Great Red

Spot, an oval-shaped area in Jupiter's southern hemisphere that has been known to exist for at least a century. Over the years, the color of this feature has changed in an irregular fashion from a brick red to a light pink, and in some years, has become almost invisible. In addition, the Great Red Spot has drifted over the planet, a fact that seems to preclude its being associated with a fixed feature on the surface. Results from the Pioneer space probes indicate that the Great Red Spot is an enormous, cyclonic storm that has been raging for centuries.

SATURN

The gas giant next distant from the sun is Saturn, whose prominent system of concentric rings make it an object of great beauty when viewed telescopically.

Motion. Saturn moves in an almost circular orbit with a radius of 9.5 AU or nearly 1.43 billion km. One sidereal period takes 29½ years to complete.

As in the case of Jupiter, Saturn's rate of rotation is best determined from the Doppler shifts present in the light reflected from the planet. By this means, Saturn has been found to possess an average rotation period of about 10½ hours, which also varies with latitude. Saturn is flattened at the poles even more than Jupiter and has an observed oblateness of about one part in ten.

Satellites. Saturn's large mass has enabled it to retain a system of ten known satellites, the largest of which is Titan, an object that, like Ganymede, is larger than the planet Mercury. Titan is known to possess an atmosphere; spectroscopic analysis has shown it to contain at least methane gas. Saturn's outermost satellite, Phoebe, is moving in a retrograde orbit and is thought to be similiar to the five outer satellites of Jupiter in its physical properties and origin.

Physical Properties. Both in size (120,800 km) and mass (95 earth masses) Saturn ranks second in the solar system. Its density, however, is the lowest of the planets and is only 0.76 that of water. Thus, Saturn would float provided a sufficiently large body of water could be made available. The interior of Saturn is thought to be similar to that of Jupiter, but no energy discrepancy such as that found for Jupiter has been observed.

Saturn's albedo (0.5) and spectroscopic analysis indicate that its atmosphere is similar to that of Jupiter, although for Saturn ammonia lines are absent, presumably because at Saturn's mean temperature

(110°K) the ammonia is frozen out. Atmospheric bands exist on Saturn but are less distinct than those on Jupiter because of the smaller size and greater distance of the former. Only a few disturbances have been observed to occur in Saturn's atmosphere, the most famous of which was the Great White Spot, a feature that appeared prominently for several months in 1933 and then dissipated. The only radio emission detected from Saturn is thermal in nature, that is, a radio emission that would be expected from a blackbody having a temperature of 110°K.

Saturn's equator and hence its ring system are tilted at an angle of

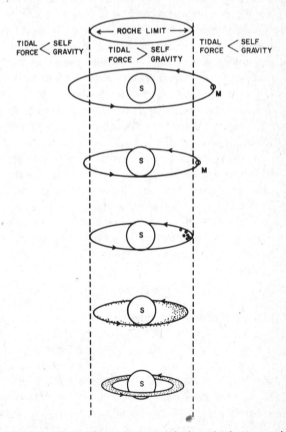

Fig. 5.7 The formation of Saturn's rings. It is thought that an ancient moon M once spiraled inside the Roche limit of Saturn, whereupon it was shattered by tidal forces. Successive collisions powdered the debris into the presently observed ring system.

28° to the ecliptic. As a result, the ring system exhibits an edge-on projection twice during Saturn's journey around the sun. At these times the rings, which are no more than a few miles thick, disappear in all but the largest telescopes. One quarter of a Saturnian sidereal period after the edge-on phase, the rings present the more familiar open phase. Measurements of the Doppler shifts present in the rings' light indicate that the various portions of the rings obey Kepler's harmonic law, which is strong evidence that the rings are made up of an infinitude of small particles, including dust, frozen gases, and meteoroid chunks, each of which moves about the planet in its own individual orbit. It is theorized that at some time in the past a satellite spiraled in close to the surface of Saturn. The tidal force exerted by Saturn on the satellite increased as the satellite distance decreased until this force exceeded the ability of the satellite to hold itself together by self-gravity. The distance at which this takes place is called the *Roche limit* for the satellite. Upon reaching the Roche limit, the satellite then disrupted along a planar symmetry as shown in Fig. 5.7, and ultimately the debris fanned out into the rings that are observed today. The gravitational action of Saturn's satellites on the particles produced gaps in the rings at distances corresponding to simple fractions of various satellites' sidereal periods, much as Jupiter has produced Kirkwood's gaps in the asteroid belt (see Chapter 7). This picture of the origin of Saturn's rings is supported by the fact that the rings are inside Roche's limit for Saturn and by the recent discovery of a tiny satellite, Janus, just beyond the outside boundary of the ring system which, like its ancient counterpart, seems to be drawing closer to Roche's limit.

URANUS

Until 1781, Saturn was the outermost of the known planets. In March of that year, the English astronomer Sir William Herschel, while conducting star counts in the constellation Gemini, came upon a small disk-shaped object that moved relative to the fixed stars on a night-to-night basis. Herschel at first thought he had discovered a comet, but further observation indicated that the object was in fact a planet. Just barely visible to the naked eye, it had previously been sighted and recorded a number of times as a star. Thus, Uranus was the first planet to be discovered by means of the telescope.

Motion. Uranus moves in a slightly elliptical orbit at a distance of 2.9 billion km from the sun and takes 84 years to complete a single revolution. Its greenish disk exhibits little more than indistinct markings and hence, the only method by which its rotational period can be measured is from the Doppler shifts present in the light reflected from the planet's edges. The value obtained by this method is 10^h49^m. One unusual aspect of Uranus's rotation is that its equator is inclined at nearly 90° to its orbital plane, thus allowing astronomers to view Uranus pole-on on some occasions and equator-on in others.

Satellites. Uranus has five known satellites, the largest of which is about 1600 km in diameter. Little else is known of these bodies.

Physical Properties. From the motions of its satellites, the mass of Uranus is found to be about 15 earth masses, and the planet exhibits a measurable disk from which a diameter of some 48,000 km is deduced.

The interior of Uranus is believed to be very much like that of the other gas giants, but the higher density of 1.5 g/cm^3 suggests a lower hydrogen abundance.

The high albedo of 0.66 for the planet indicates the presence of a deep, opaque atmosphere similar to the atmospheres of Jupiter and Saturn. Visual observers have reported faint, dusky atmospheric bands on the disk of Uranus, and spectroscopic analysis of the planet's light have established the presence of methane and hydrogen.

On March 10, 1977, during an occultation of an 8th magnitude star by Uranus, astronomers discovered a set of at least nine rings of debris surrounding the planet, ranging in diameter from 84,000 to 10,000 km. None of the Uranian rings have widths exceeding 10 km, and as such, are quite unlike the rings of Saturn, which have widths of several thousand kilometers each.

NEPTUNE

Within a decade after the discovery of Uranus, astronomers and mathematicians eagerly computed the new planet's orbit and found, much to their dismay, that Uranus did not move according to prediction, even when the gravitational effects of the planets were taken into account. By 1840, the disagreement between observation and prediction had soared to over two minutes of arc, an unacceptable discrepancy for the telescopic age. In an attempt to resolve the dilemma, two mathematicians, John Adams of England and U. Leverrier of France, simultaneously but independently predicted the existence of a new planet orbiting the sun beyond the planet Uranus whose gravitational effects were responsible for the observed discrepancy. On the night of September 23, 1846, an astronomer at Berlin Observatory found the new planet, called Neptune, within a degree of the position predicted for it.

Motion. Neptune orbits the sun in the most nearly circular orbit of all the planets. At a distance of 4.5 billion km from the sun, Neptune requires 165 years to complete a single revolution.

As in the case of Uranus, Neptune's rotational period can be measured only from the Doppler effect. The value was found to be about 16^h.

Satellites. Neptune has two known satellites, Nereid and Triton. One of these, Nereid, moves in a highly eccentric orbit that carries it from 2.1 million km out to nearly 16 million km from Neptune. With a suitable gravitational perturbation, it could escape Neptune's grasp with relative ease.

Physical Properties. In nearly every respect Neptune is a twin to Uranus. Neptune is 17 times as massive as earth and has a diameter of 45,000 km. Its mean density (2.3 g/cm^3), like that of Uranus, is higher than those of Jupiter and Saturn, again suggesting a reduced abundance of hydrogen and possibly helium in the planet's interior.

The albedo (0.62) is very close to that of Uranus, and as in the case of Uranus, methane and hydrogen have been detected spectroscopically in the Neptunian atmosphere. Thus, all available evidence indicates that the atmospheres of Uranus and Neptune are extremely similar.

PLUTO

Even after the successful prediction of the existence of the planet Neptune, the observed motions of both Uranus and Neptune failed to agree exactly with predictions. This discrepancy led several astronomers around the turn of the century to predict the existence of yet another planet beyond the orbit of Neptune. The search for the predicted planet did not yield the quick results that were obtained in the case of Neptune, and many astronomers began to doubt whether such a planet even existed. Percival Lowell, however, was particularly tenacious, and after his death the observatory he founded in Flagstaff, Arizona, carried on the search photographically until finally in 1930 a young assistant, Clyde Tombaugh, found the long sought-after planet as a 15th-magnitude object in the constellation of Gemini.

Motion. Pluto's orbit is unusual in several respects. First of all, its inclination to the ecliptic (17°) is higher by a factor of 2 than that of any other planet. Pluto's orbital eccentricity (0.25) is also the largest of the planets, and as a result, although its mean distance from the sun during its 248-year journey is about 5.9 billion km, it can come as close to the sun as 4.3 billion km or slightly closer than the planet Neptune. Because of its orbital inclination, however, the planet can never come closer to Neptune than some 384 billion km, or roughly twice the distance between the sun and Mars. There is, however, a strong possibility that at some time in the past history of the solar system, the two planets were much more closely associated and that Pluto may have been a satellite of Neptune.

Pluto is barely discernible even when viewed through the largest telescopes, and thus its rotation rate cannot be measured from markings on its surface. Moreover, its intrinsic faintness precludes the use of high-dispersion spectroscopy to measure a Doppler shift. Photoelectric measurements have shown, however, that the brightness of Pluto decreases significantly every 6.4 days, suggesting that a darker area on the planet's surface faces the earth once every rotation period.

Satellites. Pluto has one known satellite. It orbits the planet once every 6.4 days at a mean distance of about 20,000 km.

Physical Properties. Because of its remoteness, little is known about Pluto. Its failure to occult a faint star in 1965 indicates that it cannot be more than about 5900 km in diameter and may well be much smaller. Mass calculations from its gravitational effects on the other planets are highly uncertain, but from the motion of its single satellite, astronomers have determined that the mass of Pluto is approximately 0.002 that of the earth. Without a reliable diameter, the albedo of Pluto cannot be determined.

Spectroscopic analysis reveals no gases in Pluto's atmosphere, a result that is hardly surprising in view of the incredibly low (40°K) temperature estimated for the planet's surface. At such temperatures, virtually every gaseous substance would be either liquefied or frozen, save hydrogen or helium which, because of the low surface gravity of Pluto, would escape into space in a matter of a few million years. Thus, if any gaseous atmosphere ever surrounded Pluto, it most certainly lies frozen solid on the planet's surface.

In all aspects, Pluto seems to have the characteristics not of the gas giants that roam the outer reaches of the solar system, but rather of their satellites. Astronomers have thus suspected for some time that

Pluto is a runaway satellite of Neptune which at one time circled that planet in a highly eccentric orbit similar to that of Nereid and was subsequently torn away by some ancient gravitational perturbation.

OTHER PLANETS

The existence of a planet, prematurely named Vulcan, between the orbit of Mercury and the surface of the sun was once predicted, but the irregularities in the orbit of Mercury described previously can be explained using relativity theory, and intensive searches for "Vulcan" have had only negative results. It is generally agreed that if such an object exists, it can be no more than a piece of interplanetary debris less than 30 km or so in diameter.

The possibility of observing a planet beyond Pluto is somewhat more likely, but such an object would have to be a great distance from the sun and/or very small. Recently a physicist at the Lawrence Radiation Laboratory predicted the existence of a tenth planet based on anomalies in the motion of Halley's Comet. However, not only have these calculations been severly challenged on theoretical grounds, but the planet does not appear in its predicted position. The question of whether planets exist beyond the orbit of Pluto is thus still open.

OTHER SOLAR SYSTEMS

Because of its bearing on the question of life in the universe, the existence of other solar systems is a topic of continuing speculation. Unfortunately, the distances to even the nearest stars are so vast that an earth-sized planet in the vicinity of 1 AU from its local sun would be totally undetectable with present instrumentation. Several nearby stars, however, exhibit a gravitational wobble in their motion due to the presence of one or more unseen or dark companions. Masses for these objects can be estimated, and for some stars, the companion's mass is of the same order of magnitude as that of the planet Jupiter. Barnard's star, in fact, is believed to have two Jupiter-size companions revolving about it. These are the only indications that astronomers presently have for the existence of other planetary systems. Any comment beyond this enters into the realm of speculation.

THE ORIGIN OF THE PLANETARY SYSTEM

The origin of the system of planets has long been debated in the scientific community. In their text, *Solar System Astrophysics,* Brandt and Hodge list no less than 21 theories that have, at one time or other, been advanced for the origin of the planetary system. Part of the problem lies in determining what aspects of the solar system that are observed are in fact related to its formation. It is almost certain that the planar symmetry of the solar system, the uniformity of the direction of the rotations and revolutions of the planets and their satellites, and the existence of dense, compact planets near the sun and huge, tenuous planets farther away are among these factors, but others, such as the geometric progression of planetary distances called *Bode's law* (see Chapter 7) may not be significant. Out of all these theories, however, two general approaches to the problem have emerged: encounter hypotheses and nebular hypotheses.

Encounter Hypotheses. All of the encounter hypotheses have the sun either colliding with or experiencing a tidal interaction of some sort with an interloping object that pulls from the sun or itself contributes the material that goes into the formation of planets. Among the objections to such theories are that (1) the vast distances between stars and the low relative motions between our sun and the nearby stars render such encounters extremely unlikely, (2) once the material is pulled into space it is theoretically impossible to account for the fact that the material contracts into planets instead of simply dissipating into the interplanetary medium, and (3) the encounter theories do not attack the problem of the sun's origin.

Nebular Condensation Hypotheses. More in keeping with the current ideas of stellar evolution are the so-called nebular hypotheses, in which the sun and the members of the solar system are formed by the gravitational contraction of a great interstellar cloud of gas and dust. Early versions of the nebular hypothesis, in particular those of Kant (1755) and Laplace (1796), had difficulty explaining the fact that the bulk of the rotational momentum of the solar system resides in the planets rather than in the sun, nor could they account for the irregularities in planetary motion. Moreover, they could not explain why planets would be formed from gaseous clouds, which have a tendency to disperse. In 1950, G. P. Kuiper surmounted many of these difficulties with his *protoplanet* hypothesis. By assuming the appropriate combination of gas and dust and taking into account the fluid dynamics associated with turbulent clouds, Kuiper found that fragments of the solar nebula, which he called protoplanets, would be formed. These in turn would coalesce into the planets, satellites, and meteoroids of the solar system. This theory is especially attractive because it regards the solar system as just one manifestation of the physical processes by which individual stars, binary and multiple systems of stars, and other planetary systems known to exist in the vicinity of our sun all were formed.

REVIEW QUESTIONS

1. Explain why Venus and Mercury exhibit a complete set of phases. Since the moon orbits the earth and exhibits a similiar set of phases, why did Galileo conclude that Mercury and Venus orbited the sun and not the earth?
2. Why is the surface of Venus as hot as that of Mercury despite the fact that Venus is almost twice as far from the sun?
3. Describe the view of the earth-moon system that would be observed by an astronomer on Mars.
4. At favorable opposition, Mars is observed to have an angular diameter of about 28 arcseconds. What is Mars's linear diameter in kilometers? *Ans.:* 6800 km.
5. Using the orbital data from any of Jupiter's satellites listed in Appendix 3, calculate Jupiter's mass in solar masses. *Ans.:* 10^{-3} solar mass.
6. Compare and contrast the properties of the terrestrial planets with those of the gas giants.
7. How are Jupiter and Saturn alike? How do they differ?
8. Assuming that Pluto is spherical in shape, has a mass 0.002 that of the earth, and is 2400 km in diameter, calculate Pluto's density. Is your result unusual? Explain. *Ans.:* 1.5 g/cm^3.
9. How does Pluto differ from the outer planets?
10. Discuss the basic theories for the origin of the solar system.

6

The Earth and the Moon

The earth and its lone natural satellite are sometimes referred to as the twin or double planet because the moon is so large in relation to its parent body. In spite of our recent moon explorations, however, the origin of the earth-moon system is still a matter of speculation.

THE EARTH

A number of branches of science, including geology, meteorology, oceanography, and seismology, have been devoted to the study of our home planet. However, despite the impressive array of knowledge that has been discovered and deduced about the earth, many key questions regarding it remain unanswered.

Diameter. From observations of the earth's shadow during lunar eclipse, the ancient Greeks were able to deduce the spherical nature of the earth. Its diameter was first determined by the Greek mathematician Eratosthenes in 250 B.C. At noon on the first day of summer, Eratosthenes measured the angle between the sun's rays and the obelisk at Alexandria in Egypt and noted that the sun simultaneously shone down a well at Syene some 800 km away, indicating that it was at Syene's zenith. By assuming that the sun's rays reaching the earth are parallel and employing simple geometric means, Eratosthenes was

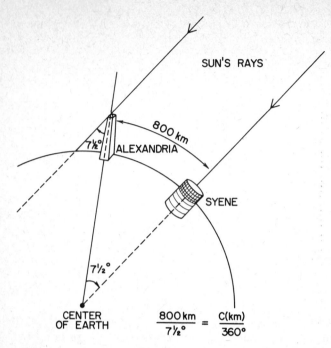

Fig. 6.1 The size of the earth by the method of Eratosthenes.

able to calculate the circumference of the earth to within a few percent (see Fig. 6.1). Extremely high-precision measurements of the earth's surface made in recent years by satellites and laser ranging techniques indicate that the earth, with some minor anomalies, is an oblate spheroid having polar and equatorial diameters of 12,714 and 12,756 km, respectively, or a flattening of about one part in three hundred.

Mass. Studies of satellite motions indicate that the earth has a mass of 6×10^{27} grams, which when combined with the earth's size yields a mean density of 5.5 g/cm^3 and a surface acceleration due to gravity of 980 cm/sec/sec or one "g."

Magnetic Field. The earth has a measurable magnetic field that behaves as if it were generated by a bar magnet buried within the earth tilted at an angle of 17° to the earth's axis of rotation.

Motion. The earth moves about the sun in a slightly elliptical orbit having a mean radius of just under 150 million km. It also rotates or spins about an axis through its center.

Proof of the Earth's Motion About the Sun. Proof of the earth's motion can be obtained by observation of stellar parallax, the radial velocity effect, the light-time effect, and the aberration of starlight. In each of these instances, none of the described physical changes would occur if the earth did not move in a closed orbit around the sun.

STELLAR PARALLAX. Stellar parallax is the apparent shift in the position of a relatively nearby star when that star is viewed against a backdrop of more distant stars from different positions in the earth's orbit.

RADIAL VELOCITY EFFECT. The radial velocity effect is a periodic variation in the observed radial velocity of a star. It arises from the earth's orbital motion, by which the earth is carried alternately toward and away from the given star.

THE LIGHT-TIME EFFECT. In the light-time effect, a regularly occurring event such as the dimming of an eclipsing binary star is observed to occur at a slightly later time than predicted when the earth is moving away from the given object and at a slightly earlier time when it is approaching it. The difference arises from the fact that the light marking the event must travel either a slightly longer or shorter distance to the earth, and hence, the detection of the event on earth is either slightly early or slightly late.

ABERRATION OF STARLIGHT. As the earth moves through space it intercepts light from the distant stars in such a way that the direction from which the light appears to come is shifted slightly in the direction of the earth's motion, much as snowflakes falling vertically appear to be moving at an angle to a car windshield as the car is driven through the storm. This effect, called the aberration of starlight, results in a back-and-forth motion of stars similar to that of parallax.

The Earth's Spin. In addition to revolving around the sun, the earth rotates or spins about an axis through its center. Intuitively, the rising and setting of stars is explained far more easily by assuming that the earth rotates beneath them than to assume that these stars revolve about the earth, thereby requiring them to traverse orbits light-years in circumference in a matter of a few hours. The Foucault pendulum, which is displayed in the lobbies of many science museums and planetariums, is often used to demonstrate the rotation of the earth. Because the pendulum is subject only to the force of the earth's gravity, no horizontal forces are acting on it and its motion should be confined to a vertical plane having a fixed orientation. Nevertheless, the Foucault pendulum will knock down in succession a series of small pegs placed in a circle about it. The only explanation for such behavior consistent with the laws of motion is to assume that the earth is rotating beneath the pendulum. In fact, it can be shown that the earth rotates beneath any object moving above its surface. To an observer on the earth's surface, the motion of such an object will appear to "break" or deflect to the right in the Northern Hemisphere and to the left in the Southern Hemisphere. This behavior is called the *Coriolis effect* and is most familiar as the agent by which the earth's cyclonic wind patterns are produced.

There are several interesting consequences of the earth's rotation, one of which is a slight degree of polar flattening. The prevailing trade winds of the middle latitudes and cyclonic storms are attributable directly to the earth's rotation. The orientation of the earth's axis of rotation is also known to wander relative to the earth's land masses within a small square-shaped area about 30 meters or so on a side. This effect, called *wandering of the poles,* is believed to be due to slight changes in the mass distribution of the earth from atmospheric motions, vulcanism, and other phenomena. There is growing evidence that the poles (or the land masses) may have moved significantly during the earth's geological history.

The axis of rotation of the earth is oriented in such a way that the equatorial plane is tilted at an angle of 23½° to the earth's orbital plane or ecliptic. As the earth moves through space, the axis of rotation maintains an almost constant orientation, and as a result, the Northern and Southern Hemispheres alternately receive the more direct rays of the sun. Thus, each hemisphere is subjected to a set of *seasons* ranging from the winter season of short days and cold weather to the summer season of long days and warm weather. Whatever season a given hemisphere is experiencing, the opposite hemisphere will be simultaneously experiencing the opposite season.

The earth's axial tilt is also responsible for the phenomenon known as the *midnight sun,* in which at high latitudes, during a given hemisphere's spring and summer seasons, the sun is constantly above the observer's horizon (see Fig. 6.2). The midnight sun can be observed in the regions between the Arctic Circle (66½° north latitude) and the North Pole and between the Antarctic Circle (66½° south latitude) and the South Pole. The polar zone not experiencing the constant light of the midnight sun has a corresponding period of darkness.

The Earth as a Clock. Because the earth's motions are strongly periodic, humans have sought to use them as a means of reckoning time. One rotation of the earth on its axis relative to the sun is called a *solar*

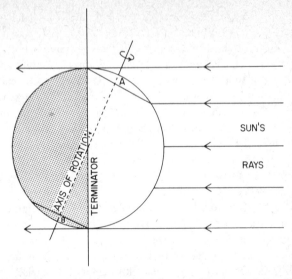

Fig. 6.2 The geometry of the midnight sun. Because of the tilt of the earth's rotational axis, an observer in zone A sees constant daylight while an observer in zone B sees only darkness.

day, and a *sidereal day,* if the rotation is measured relative to the fixed stars. As a result of the earth's orbital motion, the solar day is slightly longer (about 4 solar minutes) than the sidereal day. Because the earth also revolves about the sun, to an earth-based observer the sun appears to move in an easterly direction among the stars at a rate of just under 1° per day. Thus, the interval of time corresponding to the solar day, which uses the sun as a reference, is slightly longer (about 4 solar minutes) than the sidereal day, which uses a fixed point (vernal equinox) on the celestial sphere as a reference.

The earth's orbital motion can also be measured by the sun's apparent movement along the ecliptic. Thus, the *sidereal year* is defined as the interval of time required for the sun to complete a circuit of the celestial sphere relative to the fixed stars. The *tropical year*, on the other hand, is the interval between successive passages of the vernal equinox by the sun. Because of the westward movement of the vernal equinox (the result of the earth's precession, p. 105), the sun reaches the vernal equinox sooner than it reaches its initial position relative to the stars; thus, the tropical year is about 21 minutes shorter than the sidereal year.

Because the ecliptic is tilted at a 23½° angle to the celestial equator, the sun's declination varies sinusoidally during the course of a single year. The point at which the sun crosses the celestial equator heading from south to north is called the *vernal equinox*. The point at which the sun's declination reaches its maximum positive value is called the *summer solstice*. The point at which the sun crosses the celestial equator heading from north to south is called the *autumnal equinox,* and the point at which the sun's declination takes on maximum negative value is called the *winter solstice*. Each of the four seasons, spring, summer, autumn, and winter, formally begins when the sun reaches the points of the vernal equinox, summer solstice, autumnal equinox, and winter solstice, respectively.

The Earth's Interior. The average density of the material on the earth's surface is about 2.7 g/cm³. The overall, or mean, density of the earth, however, is 5.5 g/cm³. This discrepancy can be accounted for only by assuming the presence of much denser material in the earth's interior. Until the end of the nineteenth century, little else could be deduced concerning the state of the earth's deep interior.

The study of earthquakes has provided information about the earth's interior. Two types of shock waves, P or primary waves and S or secondary waves, emanate from an earthquake. The P waves are longitudinal, high-velocity waves that can travel through liquids as well as solids. The slower-moving S waves, on the other hand, are transverse waves (see Fig. 2.1) that are not capable of traversing a liquid medium. From the differential passage of these two types of shock waves through the earth's interior, as recorded by observing stations around the world, seismologists have been able to piece together a picture of the conditions in these regions of the earth. A schematic diagram of the earth's interior is presented in Fig. 6.3.

THE CORE. The center of the earth is occupied by a solid nickel-iron core that has a diameter of some 2700 km and is surrounded by a spherical layer of molten nickel and iron 2100 km thick. The temperature of the core is believed to be several thousand degrees Kelvin.

THE MANTLE. Surrounding the liquid core is a 2900-km-thick plastic envelope of silicon dioxide called the mantle. It is believed that the interaction of the interfaces surrounding the liquid core that result from the earth's rotation produce a dynamo effect, which in turn generates the magnetic field of the earth.

THE CRUST. The crust surrounds the mantle and ranges in thickness from about 5 km at the bottom of the oceans to over 32 km at loca-

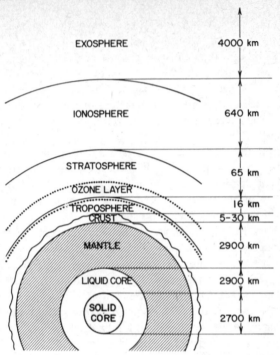

Fig. 6.3 A schematic diagram of the various layers of the earth's interior and atmosphere.

tions inland. As the outermost layer of the earth, it exhibits the effects of complicated geological forces, such as the formation of vocanos and the raising of mountain ranges. Most fascinating of all, however, is the discovery that the continental land masses have, over geological history, moved not only relative to the geographical poles, but also relative to one another. This movement, called *continental drift,* was first suspected in 1910 by the German meteorologist Alfred Wegener, who noticed that the coastlines of the various continents fit together like pieces of a gigantic jigsaw puzzle. The eastern coast of South America, for example, fits into the right-angle bend of the West African coastline that forms the Gulf of Guinea. Present evidence seems to indicate that all the continents were once joined into two enormous land masses located in the earth's tropic zones. Geologists have named these supercontinents Laurasia and Gondwanaland.

The Earth's Surface. The earth's surface is the interface between the top layer of the crust and the lowest layer of the atmosphere. The surface of the earth, like that of Mars, possesses a wide range of features that have been formed and shaped not only by geological forces from the interior, but also by various types of erosive processes on the surface itself. The age of the earth can be estimated from the degree to which certain radioactive materials in the crust have decayed into other substances. Such measurements indicate that the earth's crust was formed some 4.5 billion years ago.

The composition of the earth's surface has been accurately determined and is found to consist of 46 percent oxygen, 28 percent silicon, 8 percent aluminum, 5 percent iron, 4 percent calcium, 3 percent sodium, 3 percent potassium, 2 percent magnesium, and lesser amounts of the remaining chemical elements.

The earth's surface is unique in two important respects. First, a significant portion is covered with vast oceans that started to form about 3 to 4 billion years ago. This layer of water, known as the hydrosphere, currently covers some three-quarters of the earth's surface and has a mass of about 1/4000 that of the entire earth.

Second, the earth is the only planet known to possess life forms. The exact nature of the forces involved in the generation of life on the earth is very poorly understood, but it is believed that simple life forms first appeared in the oceans some 3 billion years ago. Since that time, life has evolved into a diversity of species and can be found in some form virtually everywhere on the earth's surface.

The Earth's Atmosphere. The earth is surrounded by a gaseous atmosphere roughly 640 km thick and composed of 78 percent nitrogen, 21 percent oxygen, and trace amounts of argon, water vapor, carbon dioxide, and other gases.

The earth's atmosphere is a constant source of frustration to the earth-based astronomer. Clouds often severely limit or destroy completely the ability to observe, atmospheric motions produce a blurring effect on the quality of celestial images, and the atmosphere's variable refractive effects wreak havoc on attempts to determine stellar positions accurately. In addition, the atoms and molecules in the atmosphere absorb, scatter, and reemit radiation, thus not only denying to the astronomer large regions of the electromagnetic spectrum, such as the ultraviolet and infrared wavelengths, but also preventing the observation of any object fainter than the level of airglow brightness. Despite such inconvenience caused by the earth's atmosphere, the astronomer's very existence would, of course, not be possible were it not for its presence.

TROPOSPHERE. The troposphere is the lowest, densest layer of the earth's atmosphere. At sea level it exerts a pressure of 15 lb/in.^{2} and has a mean density of about 10^{-3} that of water. It is in this 16-km layer that virtually all of the world's weather phenomena occur. In the upper layers of the troposphere, rapidly moving air currents known as *jet streams* encircle the globe. The jet streams are variable in nature and move at velocities as high as 320 km/hr. They are known to have a profound effect on the weather patterns closer to the earth's surface, but the exact nature of this effect and the causes for their observed variability are poorly understood.

STRATOSPHERE. Above the troposphere is a 64-km-thick layer called the stratosphere. At the base of the stratosphere conditions exist for the formation of the ozone molecule (O_3). This *ozone layer*, as it is called, is an efficient absorber of some of the dangerous shorter-wavelength radiation, especially in the ultraviolet, that impinges on the earth.

The temperature in the stratosphere remains at a constant 218°K up to an altitude of about 40 km and then begins to rise to about 265°K at the base of the ionosphere 80 km above the surface of the earth.

IONOSPHERE. The ionosphere extends 640 km above the stratosphere. In this layer of the atmosphere, short-wavelength photons and high-energy atomic particles from the sun and interplanetary medium strike the comparatively numerous atoms and molecules of the atmospheric gases there, stripping the electrons from a good percentage of the atoms. The result is the layer of charged particles or ions from which the ionosphere derives its name. The ionosphere reflects radio waves having wavelengths longer than about 20 meters, making possible worldwide radio communication.

Meteoroids are destroyed by friction in passing through the ionosphere (see Chapter 7). It is here also that *auroras* are produced. Bursts of high-energy atomic particles from the sun striking these layers of the atmosphere excite the electrons in the atoms of the atmospheric gases to higher energy levels. As these atoms deexcite, they reemit photons in the visible region that are observable as the beautiful aurorae. The earth's magnetic field lines present in the regions above the ionosphere force these particles to hit the atmosphere in the vicinity of the magnetic poles. Thus, the aurorae or northern or southern lights are most frequently observed in the high latitudes near the magnetic poles.

EXOSPHERE OR MAGNETOSPHERE. The exosphere, the last of the earth's atmospheric layers, extends from 720 km outward to over 4800 km from the earth's surface. It is characterized by a series of donut-

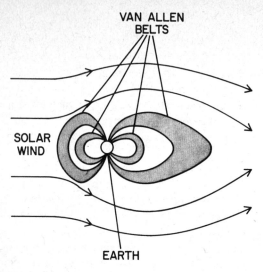

Fig. 6.4 The Van Allen belts.

shaped regions (see Fig. 6.4) called the *Van Allen belts,* in which charged particles trapped by the earth's magnetic field bounce back and forth between the magnetic poles. The sunward portions of these regions are flattened by the solar wind. The outer regions of the Van Allen layers are often referred to as the magnetosphere.

THE MOON

The earth's moon is the largest satellite of any of the terrestrial planets—Mercury, Venus, Earth, Mars, and Pluto—and rivals most of those possessed by the gas giants—Jupiter, Saturn, Uranus, and Neptune. Because the moon is relatively close to the earth, astronomers have learned more about it than any other extraterrestrial object. It was toward the moon that radar signals were first beamed, thus making it the first celestial object to be contacted by human technology. Since then, instrument-laden unmanned space probes have hit, missed, and circled the moon, and in July 1969 the Apollo astronauts touched down on its surface. The experiments set up and left behind by the Apollo astronauts and the samples of the lunar surface brought back to earth have confirmed the picture of a rugged, lifeless world that has continued relatively unchanged since the birth of the solar system.

General Properties. The mean distance to the moon has been measured by a variety of methods, including laser ranging, radar echoes, and geocentric parallax, and is found to be some 384,000 km from the earth. This value, combined with the moon's apparent angular diameter of ½° yields a linear diameter of 3460 km for the moon or about one-fourth that of the earth. The lunar mass can be obtained from its gravitational effects on the motions of the earth and from a careful study of the behavior of spacecraft, such as the lunar orbiters or the Apollo modules, when moving in the lunar gravitational field. The mass of the moon is found to be 1/81 that of the earth or about 7×10^{25} grams.

The moon's mean density is 3.3 g/cm^3, which is nearly the same as the density of its surface material and very similar to the mean density of the earth's crust. From the mass and radius of the moon, the acceleration due to gravity at the lunar surface is found to be about 5.3 m/sec/sec or one-sixth that of the earth.

Motion. The moon orbits the earth in a somewhat eccentric ellipse at a mean distance of 384,000 km. The period required to complete one such revolution relative to the fixed stars is called the *sidereal month* and is 27.3 days long, whereas a single revolution of the moon with respect to the sun is called a *synodic month* and is 29.5 days long. Because it takes the moon much less time to move about the earth than it takes the earth to move about the sun, the moon presents different aspects of its illuminated surface to the earth. The result is a series of *phases* in which the moon's visible shape changes from a slender crescent near the time of new moon to a full circle at the time of full moon (see Fig. 6.5). The synodic month is thus also equal to the period required for the moon to pass through a complete cycle of phases.

As a result of the perturbations, or small-scale effects, caused by the gravitational pull from the sun and the earth's equatorial bulge, the orientation of the lunar orbit in space changes slowly. The *perigee point,* or point of the moon's closest approach to the earth, for example, precesses in an easterly direction about 40° per year, and the points of intersection between the lunar orbit and the ecliptic or nodal points precess in a westerly direction at a rate of slightly under 20° per year.

A second consequence of the earth's gravitational pull on the moon is that the moon is gravitationally locked in on the earth in such a way that the same side of the moon is constantly facing the earth. Various

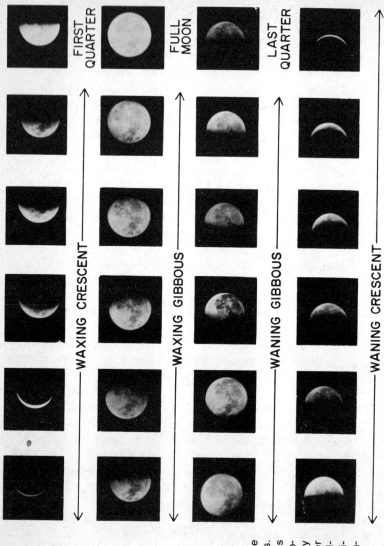

Fig. 6.5 The cycle of lunar phases. The photographs were taken at approximately 1-day intervals. (Roger B. Culver, the Colorado State University Observatory.)

geometrical and physical effects called *librations* permit earth-based astronomers to observe about 60 percent of the lunar surface. The remaining 40 percent was denied to the astronomer until recent years when lunar orbiters sent back detailed photographs.

The Moon's Interior. Seismographs placed on the moon by the Apollo astronauts indicate that moonquake activity in the lunar interior is much less frequent and severe than corresponding activity on the earth. The lack of a measurable general magnetic field on the moon and the fact that its surface density is not appreciably different from its overall density preclude a lunar version of a liquid, metallic core. The lunar orbiters and Apollo missions indicate the existence of concentrations of mass, or *mascons,* buried beneath the lunar surface. The origin of the mascons is not clear; they could be buried meteorites or volcanic plugs. Regardless of their origin, however, their existence rules out even a molten core of nonmetallic materials, because the mascons would sink to the center of the moon if the lunar interior were molten. Thus, the limited data available on the lunar interior indicate that it is a relatively cool region having no metallic component or geological activity.

The Surface of the Moon. The moon's surface presents a wealth of detail, much of which can be viewed through even a small telescope. Visible to the naked eye are the *maria,* large, relatively smooth, dark areas that are probably huge lava flows that have long since cooled. In addition to the dark areas, the naked eye can detect the much rougher, brighter areas that are sometimes called continents. Of some interest is the fact that over 60 percent of the side of the moon facing the earth is covered by the maria, whereas less than 10 percent of the extremely rugged back side of the moon has such features.

A close-up view of the moon reveals a surface riddled with *craters.* These craters range in size from depressions fractions of an inch across to gigantic *walled plains* over 160 km in diameter, which are believed to be ancient impact craters that have since filled up with lava. Some of these craters exhibit systems of rays that extend dozens of miles across the lunar surface, strongly suggesting that the given crater was formed by an impact of tremendous magnitude.

The lunar landscape is also dotted with a large number of *mountain ranges* and individual peaks. From the lengths of the shadows cast by these features, astronomers and space technicians have determined that these mountains often reach 6000 meters or more above the surrounding terrain. Winding their way through the mountains and craters like ancient river beds are the channel-like depressions called *rilles*. Despite their appearance, it is reasonably certain that the rilles are not due to the action of flowing water, but more likely have arisen from lava flows or possibly venting of gases from the lunar interior. Fault lines, such as the Straight Wall (Fig. 6.6), similar to those observed on earth have also been found at several places on the lunar surface. Other regions are sprinkled with humped features called *domes*.

Exploration of the lunar surface in a few selected sites by Apollo astronauts has given astronomers and geologists only a tantalizing close-up glimpse of the moon. From these glimpses, the moon appears to be a completely lifeless body coated with a glassy layer of dust a few inches thick, which is believed to be the result of millions of years of meteoric impacts. Analyses of the samples of lunar soil and rocks brought back to the earth indicate that whereas the moon's crust has roughly the same relative composition of elements as the earth's, some mineral combinations of these elements are unique to the lunar surface. The lunar material is also found to be much older than anything found on the earth. Dating of lunar samples by geologists indicates that some rocks and grains of dust are as much as 4.7 billion years old, or very nearly as old as the solar system itself.

It is generally agreed that the lunar surface has been shaped and formed by both meteoric impact processes (craters, ray systems, etc.) and volcanic processes (maria, fault lines, domes, etc.). The limited studies of the lunar interior, however, point to the idea that the moon is now geologically dead and that volcanic processes, although they played a significant role in shaping the lunar surface in the moon's younger days, have virtually no effect at present. Because of the lack of water or air on the moon's surface, the moon has not been subjected to the wind, rain, and running water that have been major factors in shaping the earth's surface.

The Moon's Lack of Atmosphere. Astronomers have long been aware that the moon lacked a significant atmosphere. The low lunar albedo, the lack of any spectral lines other than those of the sun in the lunar spectrum, the lack of dimming or extinction of the brightnesses of objects about to be blotted out or *occulted* by the edge of the moon's disk, and the low velocity required for gases to escape the moon's surface all pointed to the fact that the moon possessed no atmosphere long before Apollo astronauts verified it.

Origin of the Moon. Three basic theories for the origin of the moon have been advanced by astronomers: (1) the moon was torn

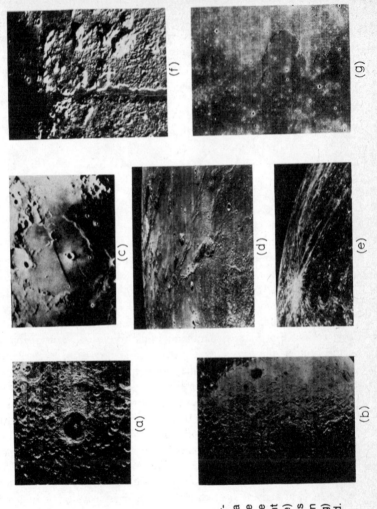

Fig. 6.6 Some lunar surface features: (a) a lava-filled crater, (b) the "bulls-eye" feature Mare Orientale, (c) the Straight Wall, (d) lunar domes, (e) an impact crater and its ray system, (f) the rill in the Alpine Valley, and (g) a lunar crater field. (NASA.)

from the earth, probably in the vicinity of the present Pacific Ocean, shortly after the formation of the earth; (2) the moon is an interloper from another region of the solar system that has been captured by the earth's gravitational field; and (3) the earth and moon were formed as a "twin planet" and that subsequent differences in the two bodies have arisen from the inability of the moon to retain either oceans or an atmosphere. The data from the Apollo missions tend to disprove the theory that the moon was torn from the earth, but the moon's origin is still very much unsettled.

EARTH-MOON RELATIONSHIPS

In terms of size and mass, the moon is larger, relative to its primary, than any other satellite in the solar system. As a result, the moon exerts a number of significant effects on the earth.

Tides. Most important of the earth-moon relationships is the raising of tides in the earth's oceans. Because the force of gravity on a particular mass varies inversely with the square of the distance of the attracting mass, the moon pulls on various sections of the earth in varying degrees depending on the distance between the section and the moon, as shown in Fig. 6.7a. If the moon's tidal forces are viewed from the earth's center, the forces appear as shown in Fig. 6.7b. Each of these forces possesses components that are perpendicular and tangential to the earth's surface as shown in Fig. 6.7c. Now the earth's surface is more rigid than steel and hence responds very little to this tidal distortion. The water at the surface of the earth, however, responds to the tangential components of the tidal force at the earth's surface and piles up into two tidal bulges, one on the side of the earth nearest the moon and the other on the side farthest away. As the earth spins on its axis, the deep and shallow regions of the water or the high and low tides will appear at a given location about twice every rotation period of the earth.

The sun exerts a similar but smaller set of tidal forces over the earth's surface. When the earth, moon, and sun are aligned, as at the time of new moon or full moon, the tidal bulges produced by the sun and moon reenforce one another and the highest tides, called *spring tides,* occur. At the time of either quarter moon, the tidal bulges produced by the sun and moon are at right angles to one another and the water depth is fairly evenly distributed over the earth's surface. These tides, called *neap tides,* are the lowest possible tides that can occur. The exact amount that a tide is raised at a given location also

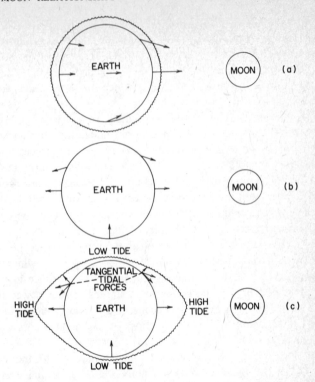

Fig. 6.7 The formation of tides. The differential lunar gravitational forces (a) when viewed from the center of the earth take on the appearance shown in (b). The tangential components of these forces in turn pile up the earth's water into regions of high and low tides as shown in (c).

depends on the geography of the region. Thus, in the Bay of Fundy in eastern Canada, which is a fairly confined body of water, tides of 50 feet or more can occur, whereas in the watery expanse of the middle of the ocean, the tides will amount to only 2 or 3 feet.

Precession. In addition to producing the tides, the gravitational pull of the moon exerts an effect on the earth's motion called precession. As previously noted, the earth's equatorial bulge is tilted at an angle of about 23½° to its orbital plane. The moon's gravitational field attempts to "straighten up" the earth's axis of rotation, but because of the earth's rotation rate, the response of the rotation axis is to slowly change its orientation, or precess, much as a tilted, rapidly spinning top will alter its axis of rotation. The sun produces a similar effect, and the combined precession resulting from the action of the sun and moon, called *lunisolar precession,* causes the north celestial pole to move among the northern stars in a circular path 47° in diameter once every 26,000 years. The difference in orbital planes of the moon and the earth, the varying distances of the sun and moon, and the gravitational action of the other planets all produce slight departures from this circular motion. The combined effect of the first two factors is referred to as *nutation;* the latter effect is called *planetary precession.* The combined effects of lunisolar precession, nutation, and planetary precession is known as *general precession.* The general precession of the north celestial pole relative to the fixed stars also causes the vernal equinox point to move along the ecliptic at a rate of 50 arcseconds per year. This motion in turn causes the celestial position coordinates of a given object to gradually change in time and thus must be corrected for whenever a celestial position is calculated.

Eclipses. As the earth and moon move through space their solid, spherical shapes generate shadow cones (see Fig. 6.8), which consist of a dark umbra and a somewhat lighter penumbra. Once in a while, the earth, new moon, and sun become aligned in such a way that the earth passes through the shadow cone of the moon. Such an event is called an eclipse of the sun. An eclipse of the moon occurs when the full moon passes into the shadow of the earth.

Because the moon's orbit is inclined about 5° to the ecliptic, eclipses of the sun and moon do not occur at each and every new

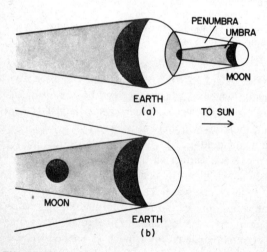

Fig. 6.8 The alignment of the earth, moon, and sun during (a) a solar eclipse and (b) a lunar eclipse.

moon or full moon, but only when the sun and moon simultaneously arrive at one of the intersection points or nodes between the lunar orbit and the ecliptic. If the arrival occurs at the same nodal point, a solar eclipse will occur, and if the arrival occurs at opposite nodal points, a lunar eclipse will occur. The number, type, and duration of possible eclipses in any given year depend on a wide variety of factors, including the orientation of the lunar orbit and the angular diameters of the sun and the moon. The conditions that lead to a particular kind of eclipse, however, are periodic in nature, and will thus repeat themselves in various cycles or families of eclipses that all have very nearly the same characteristics. The most famous of these eclipse cycles is the *saros cycle,* which was known to a number of the ancient cultures. In this cycle, similar solar eclipses will occur successively about once every 18 years.

Solar Eclipses. For observers located at any spot in the lunar shadow cone (see Fig. 6.9), the sun will appear to be blotted out, or eclipsed, in varying degrees. Observers located in the umbra of the moon's shadow see a *total eclipse* of the sun, whereas those in the penumbra will see a *partial eclipse.* In some instances the convergent point of the moon's shadow cone fails to reach the surface of the earth, and an *annular* or donut-shaped eclipse is observed.

During the early stages of a total solar eclipse, the dark, curved edge of the lunar disk appears to engulf more and more of the sun's disk. As the last rays of sunlight are blotted out, they appear to peek out from the edge of the moon, giving rise to the so-called *diamond ring effect.* This single "gemstone" of light is then quickly replaced by a phenomenon known as *Bailey's beads,* a series of flowing bright spots that result from the last vestiges of sunlight shining through the gaps in the rugged topography at the edge of the lunar disk. Moments later, the entire solar disk is covered by the moon, and the lunar disk is surrounded by the soft glow of the solar *corona,* the outermost layer of the sun's atmosphere. During this total phase, or *totality,* the corona shines with the light of the full moon. After a few minutes, these events reverse themselves as the sun reemerges from behind the lunar disk.

During a total solar eclipse, the sun's visible disk, or *photosphere,* is blotted out by the moon, permitting the astronomer to study the less luminous outer layers of the solar atmosphere, especially the corona. In addition, the properties of the upper layers of the earth's atmosphere, in particular the ionosphere, can be observed during conditions of sudden temperature drops, thus affording a better understanding of

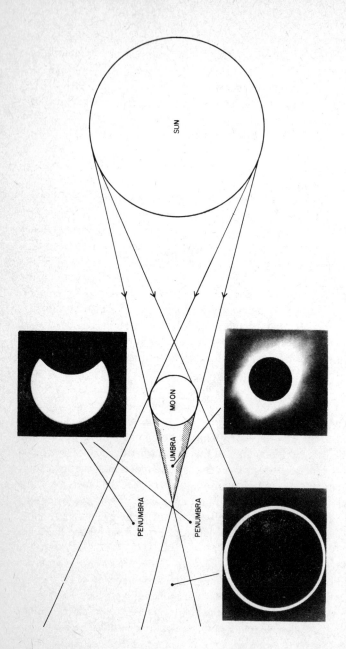

Fig. 6.9 The lunar shadow. An observer situated at each of the given total and partial eclipse points in the shadow will observe the eclipse indicated. (Roger B. Culver, Colorado State University Observatory; Annular Eclipse, Griffith Observatory.)

these regions above the earth. Most intriguing of all, perhaps, is the possibility that the gravitational bending of starlight predicted by relativity theory for star images in the vicinity of the sun can be measured during the total phase of a solar eclipse. Attempts to measure it have indicated that the bending is indeed present, but have not yielded a very accurate value for the deflection.

Eclipse paths, or zones on the earth within which a particular solar eclipse may be observed as total, are seldom over a few dozen miles wide or longer than a few hundred miles. Even in the path of totality under the most favorable conditions, an observer in a fixed location can view only about 7½ minutes of totality, and in most instances, the length of totality is much shorter. In recent years, astronomers have sought to increase the observed length of totality by flying above any cloud cover and racing the moon's shadow down the eclipse path in high-altitude, high-velocity jet aircraft, thus gaining more time in which to perform their observations and experiments.

Lunar Eclipses. Lunar eclipses are either partial or total, never annular. During a total lunar eclipse, the curved shadow of the earth slowly creeps across the face of the full moon. At total phase, however, a certain amount of sunlight is refracted around the edges of the earth by the terrestrial atmosphere, causing the moon to glow a dull coppery color. Unlike solar eclipses, lunar eclipses are of little use to the astronomer, save for the precise determination of the positions of the earth and moon relative to the sun.

REVIEW QUESTIONS

1. Describe the interior of the earth. How is this region of the earth investigated?
2. How does the earth's surface differ from that of the other planets?
3. Compare and contrast the various layers of the earth's atmosphere.
4. Why does the moon exhibit a series of phases?
5. How is the lunar interior different from that of the earth?
6. Compare and contrast the processes that have formed and altered the surface of the earth with those that have formed and altered the lunar surface.
7. What have the various space missions discovered about the moon that was not previously known?
8. Discuss the various theories for the origin of the moon.
9. Describe the ways that the moon affects the earth.

7

The Interplanetary Medium

The first evidence that the space about the sun and planets is not a complete void was the observation of meteorite falls. These "stones from heaven" were reported at various times from the seventh century B.C., but it wasn't until early in the nineteenth century that their extra-terrestrial nature was clearly recognized. "Falling" or "shooting" stars, now called meteors, were also observed. Even more spectacular were the comets, apparitions that appeared in the sky, blossomed for a few days or weeks, and then disappeared, leaving a trail of gaseous, dusty debris. Early in this century it was also discovered that the earth is subjected to a flux of ionizing particle radiation, the cosmic rays, which originate outside of the earth's atmosphere. From all of these observations a picture has evolved of a vastly complex interplanetary medium consisting of gas, dust, and chunks of stone and metallic material as large as several hundred kilometers in diameter.

THE ASTEROIDS

Lesser bodies of the solar system are not distributed throughout the interplanetary medium but are concentrated in the region between the orbits of Mars and Jupiter called the *asteroid belt*. The vast array of meteoroid chunks that have all or part of their orbital paths in this region are known as *asteroids*.

Discovery of Asteroids. In 1766 the mathematician Johannes Titius formulated the following scheme for remembering the distances to the planets: Set down the progression 0, 3, 6, 12, and so on, in which the next element of the progression after 3 is obtained by doubling the previous one. By adding 4 to each element and then dividing the result by 10, one obtains the approximate distances of each of the planets from the sun in astronomical units (see Table 7.1). This progression is called *Bode's law* after Johann Bode, who made frequent reference to it. However, it is not a law, but only a useful relationship, for it predicted the existence of Uranus but not of Neptune. It also indicated the existence of a planet between the orbits of Mars and Jupiter. The discovery of Uranus in 1781 spurred astronomers to search the skies for the missing planet. On the night of January 1, 1801, a Sicilian astronomer named Giuseppe Piazzi accidentally discovered a planet-like object circling the sun at the predicted 2.8 AU. This new object was called Ceres and was believed to be the planet they sought. Much to the surprise of astronomers, however, a second asteroid, Pallas, was discovered a little over a year after Piazzi's first observations of Ceres. Two more asteroids, Juno in 1804 and Vesta in 1807, were discovered before the end of the decade. By the end of the nineteenth century the total number of known asteroids was well over 300, and at present some 1700 asteroids have reasonably well-determined orbits. It is estimated that tens of thousands of asteroids are within the reach of the world's great telescopes.

Table 7.1. Bode's "Law" and the Corresponding Planetary Distances

Progression Element	Distance to Sun from Bode's Law (AU)	Planet Corresponding to Progression Element	Actual Planet-Sun Distance (AU)
1	$^1/_{10} (4 + 0) = 0.4$	Mercury	0.39
2	$^1/_{10} (4 + 3) = 0.77$	Venus	0.72
3	$^1/_{10} (4 + 6) = 1.0$	Earth	1.0
4	$^1/_{10} (4 + 12) = 1.6$	Mars	1.6
5	$^1/_{10} (4 + 24) = 2.8$	Asteroids	2.8 (mean)
6	$^1/_{10} (4 + 48) = 5.2$	Jupiter	5.2
7	$^1/_{10} (4 + 96) = 10.0$	Saturn	9.5
8	$^1/_{10} (4 + 192) = 19.6$	Uranus	19.2
9	$^1/_{10} (4 + 384) = 38.8$	Neptune	30.6
10	$^1/_{10} (4 + 768) = 77.2$	Pluto	39.4

Physical Properties of Asteroids. Of all the known asteroids, only four have measurable angular diameters and hence linear diameters that can be determined directly. These are Ceres (700 km), Pallas (460 km), Vesta (380 km), and Juno (220 km). The sizes of the remaining asteroids are estimated from their apparent brightnesses by assuming that their albedos are about 0.1, the observed average albedo for the four asteroids mentioned above. An asteroid's apparent brightness, then, depends on its distance from the sun, its distance from the earth, and its cross-sectional area. The cross-sectional area of the asteroid is equal to $\pi \times$ (radius of asteroid)^2, and the radius and diameter of a given asteroid can easily be calculated if the earth-asteroid and the sun-asteroid distances are known. Astronomers have found that there are about fifteen asteroids with diameters greater than 160 km, 400 or so with diameters greater than 16 km, and several thousand with diameters greater than 2 km.

Asteroid masses are roughly estimated by assigning to them the density of stony meteoric material (about 3 g/cm^3) and multiplying this density by the volume for a sphere or whatever other shape the given asteroid may have. Using this method, asteroid masses are found to range downward from about 10^{-4} earth mass, and the total mass of all the asteroids is estimated to be about $1/1600$ that of the earth.

A great many asteroids are irregular in shape, as evidenced by the large variations in their light curves. Systematic photometric observations of the asteroids reveal that these objects come in a wide variety of sizes and shapes and rotate with periods ranging from 2 to 20 hours.

Orbits of Asteroids. Because asteroids are low-mass objects, it is relatively easy for them to be perturbed out of a given orbit by the gravitational action of the planets, especially Jupiter. Thus, although most of the asteroid orbits are ellipses that lie between the orbits of Mars and Jupiter and are reasonably well confined to the plane of the solar system, others exhibit some highly interesting effects. Some asteroids, such as the Trojans, are gravitationally trapped into a fixed position relative to the sun and Jupiter and are carried ahead and behind the gas giant as it orbits the sun. Others seem to be even more strongly trapped by Jupiter's gravity and are now orbiting Jupiter as satellites. Still others travel great distances away from the asteroid belt. The path of the asteroid Hidalgo reaches out to the orbit of Saturn, and that of Icarus passes within 32 million km of the sun, well within Mercury's orbit. On occasion, some asteroids pass very close to the earth and in the earth's recent geological history, have actually struck our planet. It is surmised, for example, that Meteor Crater (Barringer Crater) in Arizona was formed by the impact of one such asteroid between 10,000 and 100,000 years ago.

The repeated gravitational action of Jupiter on the asteroids has cleared a series of zones within the asteroid belt called *Kirkwood's gaps*. The gaps are located at mean distances that, from Kepler's harmonic law, correspond to sidereal periods that are even fractions such as one-half, one-third, or one-fourth of Jupiter's sidereal period. These gaps are analogous to the divisions observed in Saturn's rings.

Origin of Asteroids. There are several theories concerning the origin of the asteroids. One theory is that they are remnants of one or more planetoids about the size of the moon that were fragmented either by the tidal action of Jupiter's gravity or by collisions with one another. Another theory speculates that the asteroids are the fragments of a planet that for some reason was unable to form along with the other planets of the solar system.

Asteroid or Comet? In November 1977, Charles Kowal of the Hale Observatories discovered an unusual object orbiting the sun with a 50.7-year period along an orbit whose distance from the sun ranges from 1.3 to 2.8 billion km. Thus, the new object, called Chiron, is closer to the sun than Saturn at perihelion and almost as far as Uranus at aphelion. The estimated diameter of Chiron is between 160 and 640 km. Because comets and asteroids are indistinguishable on the basis of appearance at this distance from the sun, it is not clear at this writing whether Chiron should be classified as a comet or an asteroid.

COMETS

Because of their spectacular and sudden appearance, comets were, for centuries, regarded as omens of evil and catastrophic events. Aristotle claimed that comets were atmospheric phenomena similar to aurorae and shooting stars. In 1577, however, Tycho Brahe carefully observed a bright comet, and failing to detect a diurnal parallax for this object, concluded that it was much more distant than the moon and was hence an interplanetary object. The interplanetary nature of the comets was established a century later when Edmund Halley, a close friend of Isaac Newton, used Newton's laws of motion to describe cometary orbits and even successfully predicted the return of a bright comet that now bears his name.

Since Halley's time, astronomers have found that virtually all comets move in highly flattened elliptical orbits whose aphelion points can range up to thousands of astronomical units from the sun. In addition, the orbits are randomly oriented with respect to the sun, so that every direction of approach for a given comet is equally probable. If comets were interstellar in nature, they would not move in closed ellipses, but in parabolic or hyperbolic encounter orbits and would enter the solar system primarily from the direction toward which the sun is moving in space. Thus, astronomers have concluded that comets are a part of the solar system.

Observed Structure of Comets. Comets are discovered at a rate of about five or six per year. Most of them are faint, telescopic objects; prominent comets visible to the naked eye appear only about once in a decade. Comet heads consist of a diffuse *coma* which may contain a bright, star-like *nucleus*. Satellite observations of the two bright comets of 1970 indicate that comet heads are also surrounded by enormous spheres of hydrogren gas up to 10 million km in diameter. The heads of comets can range up to several hundred thousand kilometers in diameter. Most comets develop tails as they approach the sun. Tails have been observed that extend over 160 million km in space.

Mass of Comets. Despite their great size, comets are objects of extremely low mass. Attempts to measure their gravitational effects on the orbiting planets of the solar system have permitted astronomers only to place upper limits to their masses. Comets are certainly no more massive than 10^{-7} earth mass and are probably on the order of 10^{-9} to 10^{-12} earth mass. Furthermore, there are several instances of comets transiting the solar disk with no observable trace, which strongly suggests that they are made up of a multitude of particles, each of which is not more than a few kilometers across.

Spectra of Comets. Although spectra vary considerably from comet to comet, several basic stages in cometary spectra have been observed in the last few decades. Spectra taken of comets at great distances from the sun reveal that the light emitted by the comet is essentially reflected sunlight, indicating that at these distances the comet contains mostly solid dust and ice particles and little or no gaseous material. At distances of less than about 3 AU, comets display spectra with emission lines of CN, C_2, OH, and NH. These molecules are fragments of the larger molecules of methane, ammonia, and water that have been excited by the sun's ultraviolet radiation and dissociated. If the comet is a ''sun grazer'' and reaches the outer limits of the solar atmosphere,

Fig. 7.1 Comet Mrkos showing a dust tail (curved portion) and a gas tail (straight portion). (Courtesy of the Hale Observatories.)

the excitation of the atoms in the comet is much more pronounced, and bright lines of single elements such as silicon, sodium, and iron appear. Some comet tails exhibit a reflected solar spectrum and are thus dust tails. Others exhibit various emission lines that suggest a predominance of gaseous material. Still others have a composite emission-line and reflected spectrum, which indicates a tail composed of both gas and dust. A comet (see Fig. 7.1) may have a system of two or more tails, some of which are gaseous and others of which are dust.

Life Cycle of Comets. It is generally believed that comets originate from a *comet cloud,* a spherical shell composed of the remnants of the material out of which the sun and planets were formed, which extends from 50,000 to 150,000 AU from the sun. A comet begins its life when some of this gaseous material is slowed down, perhaps by the attraction of a passing star, and falls toward the sun. During its fall, the gaseous chunk acts like an interplanetary vacuum cleaner and gathers up large amounts of frozen gas, dust particles, and ices. When this collection of matter reaches the inner portions of the solar system, the frozen ices and gas sublime to form a glowing atmosphere (the coma) about the particle swarm (the nucleus). As a comet approaches the sun, some of the coma is forced out of the head by the solar wind and by radiation pressure into a luminous tail which grows in length as the distance between the comet and the sun decreases. Caught up in the solar wind, the comet tail always points radially away from the sun.

As a comet leaves the sun, the tail grows less luminous and begins to dissipate. In some instances a comet is slowed down by an encounter with one of the gas giants and falls back toward the sun in a much smaller orbit, its aphelion point being in the vicinity of the retarding planet's orbit. Such a comet is called a *periodic comet* and is said to belong to that planet's family of comets. Jupiter, for example, has collected more than forty such comets.

After several hundred perihelion passages the material lost to a comet from the formation of the coma and tail reaches a significant fraction of its total mass, and the comet begins to disintegrate. For example, Biela's comet, with a period of 7 years, was observed to split in two in 1846. Both components returned in 1852. Since that time, however, Comet Biela has not been seen. Once a comet disintegrates, the remnants move along the comet's orbit as a meteoroid swarm which gradually dissipates into the interplanetary medium.

METEORS AND METEORITES

As the earth moves through space, it encounters a great many solid particles called *meteoroids*. Any meteoroid that happens to be in the earth's way will be drawn toward it by the force of gravity. The friction between the rough surface of the rapidly moving (12–72 km/sec) meteoroid and the molecules in the earth's atmosphere heats up the meteoroid to several thousand degrees and causes it to vaporize. The light from this event is observed as a *falling star, shooting star,* or *meteor*. The brightness of meteors can vary considerably with the velocity of the incoming meteoroid relative to the earth's atmosphere, with the mass of the meteoroid, and with the meteoroid's composition. When a very energetic interaction occurs between a meteoroid and the earth's atmosphere, the resulting meteor will be particularly bright, often attaining the brilliance of a full moon. Such meteors are called *fireballs*. *Bolides* are meteors that explode with an audible sound.

Meteor Showers. Periodically the earth's orbital motion brings it into contact with the remnants of disintegrated comets. When this occurs, the earth's atmosphere is bombarded with thousands of particles, each of which is visible as a meteor. Such a display of meteors is referred to as a *meteor shower*. Some meteor showers rank among the most spectacular of all the celestial events. Large meteor showers occurred, for example, when the earth encountered the remnants of the ill-fated Comet Biela in 1872 and again in 1885. More recently, impressive displays of Leonid meteors occured in November 1966 and of Draconid meteors in October 1946. Such displays, however, are com-

Table 7.2. Some Annual Meteor Showers

Shower	Date of Maximum Number of Meteors	Usual Number of Meteors per Hour
Quadrantids	Jan. 3	30
Lyrids	Apr. 21	15
Perseids	Aug. 11	50
Draconids	Oct. 9	10
Orionids	Oct. 20	15
Taurids	Oct. 31	10
Leonids	Nov. 16	15
Geminids	Dec. 13	50

paratively rare. The cometary debris usually encountered by the earth is distributed fairly evenly along the comet's orbit, with the result that the shower occurs regularly each year, but with fewer meteors per unit time. The most reliable meteor showers of this type are the Perseids in August and the Geminids in December (see Table 7.2).

The particles giving rise to meteor showers move parallel to one another in space. However, as they strike the earth's atmosphere they appear to diverge from a point called the shower *radiant,* much as a pair of parallel rails seems to radiate from a distant point (see Fig. 7.2). The shower usually takes its name from the constellation in which this radiant point is located. For example, the Leonids appear to radiate from the center of the sickle of the constellation Leo.

Sporadic Meteors. Meteors that do not occur in showers are called sporadic meteors because they appear at random and do not seem to radiate from any single point. Moreover, their orbital paths prior to entry into the earth's atmosphere are randomly oriented in space.

Meteorites. If conditions are right, a sporadic meteoroid, or part of

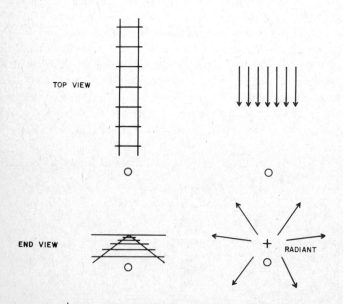

Fig. 7.2 The geometry of a radiant point. Because of a perspective effect, a set of particles moving parallel to one another will appear to the end-on observer 0 to diverge from a radiant point, just as in the more familiar case of a pair of parallel track rails.

Fig. 7.3 The Widmanstätten figures. (Roger B. Culver, the Colorado State University Observatory.)

one, can successfully traverse the earth's atmosphere and strike the surface. Such a chunk of interplanetary material is called a *meteorite*. Meteorites are generally grouped into three classes, the irons (sideri-tes), the stones (aerolites), and the stony-irons (siderolites). *Iron mete-orites* consist of approximately 90 percent iron, 9 percent nickel, and 1 percent other substances, a composition not dissimilar to that thought to exist in the earth's core. When sliced and polished and etched with acid the irons exhibit a unique crystalline structure known as the Widmanstätten figures (see Fig. 7.3). *Stony meteorites* have a pre-dominately silicate composition much like the earth's crust and as such, their identification as meteorites is much more difficult than for the irons. Rarest of all the meteorites are the *stony-irons*, which are composed of a mixture of almost equal amounts of iron and stony meteorite material.

In 1969, eight amino acids with a slightly different chemical struc-ture from those found on earth were detected in a group of meteorites that fell near Murchison, Australia. This remarkable discovery suggests the existence of extraterrestrial processes capable of building the com-plicated molecules needed for life as we know it.

Meteorites can reach considerable size. The largest known meteorite lies in Southwest Africa and weighs some 60 metric tons. Throughout

Fig. 7.4 Meteor Crater (Barringer Crater) in northern Arizona. (Yerkes Observatory.)

the world meteoritic craters of considerable size, such as the one in northern Arizona (see Fig. 7.4), bear witness to the fact that every so often the earth is struck with enormous impact by meteorites hundreds of meters across.

Round, glassy objects called *tektites* have been found in several regions of the earth's surface. Their composition is not like that of the meteorites, and their true nature and origin are a matter of controversey.

INTERPLANETARY DUST

Tiny dust-like particles have been observed and collected in the upper atmosphere by balloons, rockets, and aircraft and in the interplanetary medium by deep-space probes (see Fig. 7.5). These particles, only a few microns in diameter (1 micron = 10^{-4} cm), seem to be composed of the same materials as meteorites, revolve around the sun in individual orbits, and have a density of about 200 particles per cubic kilometer.

Micrometeorites. When interplanetary dust particles come into contact with the earth's atmosphere, they are decelerated and thus do not heat up and vaporize as a meteor does. Rather, they float to the surface of the earth comparatively unaltered, much as dust settles after a wind storm. These particles, too small to be observed as meteors, are referred to as micrometeorites.

Zodiacal Light. On a dark, clear night it is sometimes possible to observe a faint band of light near the horizon. This band, called the zodiacal light, is symmetrical with respect to the ecliptic and increases in intensity with decreasing angular distance from the sun (see Fig. 7.6). The zodiacal light is thought to be due to the scattering of sunlight by interplanetary particles.

The Gegenschein. The gegenschein, or counterglow, is a faint, diffuse patch of light that can be seen almost exactly opposite the sun on a dark, clear night. Like the zodiacal light, the gegenschein is believed to be due to the reflection of sunlight by interplanetary dust particles.

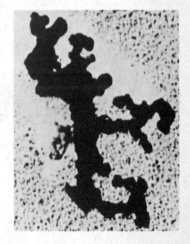

Fig. 7.5 Two micrometeorites. The width of each particle is about 10^{-4} cm or 1 micron. (Courtesy C. L. Hemenway, Dudley Observatory.)

Fig. 7.6 An artist's conception of the zodiacal light. Because of the nature of the zodiacal light, only an artist's hand can convey its appearance to the naked eye. (Bonnie Culver.)

INTERPLANETARY GAS

There is evidence that gas as well as dust exists in the interplanetary medium. This gas is composed primarily of protons and electrons and has a density of 1 to 5 particles per cubic centimeter. Solar phenomena, including flares, coronal streamers, and the solar wind (see Chapter 8) are believed to be the source of most of the interplanetary gas. Other high-energy particles found in interplanetary space may originate within our galaxy from supernovae explosions or flare-type particle ejections from distant stars. The high-energy particles emanating from the sun as well as from outside the solar system are often referred to as *cosmic rays* (see Chapter 11).

REVIEW QUESTIONS

1. Describe the physical characteristics of a comet.
2. Summarize a comet's life cycle.
3. Compare and contrast the following: meteor, meteoroid, meteor shower, meteorite, fireball, micrometeorite.
4. Describe the various types of meteorites.
5. Asteroid A is five magnitudes fainter than asteroid B. If both are the same distance from the earth and the sun and the diameter of asteroid B is 64 km, what is the diameter of asteroid A? *Ans.:* 6.4 km.
6. An asteroid has a diameter of 2×10^5 cm. If the density of this asteroid is assumed to be 3 g/cm^3, what is its mass? *Ans.:* 1.3×10^{16} g.
7. What is the zodiacal light? The gegenschein?
8. What are cosmic rays?
9. Calculate the distance in astronomical units between Pluto and the sun from Bode's law. How does this value compare with the actual value of 39.4 AU? *Ans.:* 77.2 AU from Bode's law.

8

The Sun

As the center of the solar system, the sun is not only larger and more massive than all the planets combined but is also the source of the light and energy by which the planets are illuminated. In the last century it was recognized that the sun is the nearest example of a distinct class of celestial objects that we see in the night sky as the stars, and that are essentially balls of incandescent gas whose energy is generated by large-scale nuclear processes.

GENERAL PROPERTIES OF THE SUN

Solar Diameter. The sun exhibits an angular diameter of about 30 seconds of arc as seen from the earth. Since the earth-sun distance is known, the linear diameter of the sun can be obtained; it is 13.92×10^5 km.

Solar Mass. The motion of the earth about the sun provides the most accurate means for determining the solar mass. By using Kepler's harmonic law expressed in centimeter-gram-seconds (CGS) units, the solar mass is found to be 2×10^{33} grams or 330,000 times that of the earth.

Solar Luminosity. The amount of energy striking the surface of the earth per unit area per unit time is called the *solar constant* and has a measured value of 1.4×10^6 ergs/sec/cm². If the solar constant is multiplied by the total surface area of a sphere having a radius equal to the earth-sun distance, then the product is equal to the solar luminosity, $L_\odot$, or the total energy generated by the sun per unit time. The solar luminosity thus determined is 4×10^{33} ergs/sec.

Solar Temperature. The mean temperature of the sun's outer layers can be estimated in a variety of ways, including the use of Wien's law, Stefan's law, Planck's law, and the relative intensities of absorption lines in the solar atmosphere. In all cases, the mean temperature of the sun's atmosphere is found to be in the vicinity of 5800°K.

Rotation of the Sun. The sun, like the planets, spins on its axis. The rate of rotation is measured by observing either the rate at which sunspots move across the visible face of the sun or from the Doppler shifts present in the light radiating from its edges. Results from such studies indicate that the sun does not rotate as a rigid body, but spins faster at its equator (once every 25 days) than at points near its poles (once every 35 days).

Solar Magnetic Field. The sun possesses a very weak magnetic field which is thought to be similiar to that of the earth, both in terms of the general shape of its field lines and its overall magnitude. The relationship between this general field and the stronger, more localized magnetic fields that are in part responsible for sunspots, flares, and other aspects of solar activity is not yet clear.

THE SOLAR INTERIOR

The sun has been shining at or near its present rate for approximately 4 billion years. Its tremendous energy output cannot be accounted for by any classical method of generation such as chemical, gravational, or rotational. The dynamics of the sun's energy generation, as well as that of the distant stars, puzzled astronomers and physicists into the early part of this century, at which time Albert Einstein demonstrated in his theory of special relativity that matter and energy are related by the familiar equation $E = mc^2$, where c is the speed of light. The conversion of even a small amount of material into its equivalent energy can result in an energy release sufficiently large to account for the intensity and duration of solar luminosity. Nuclear energy is released either by breaking apart atoms (*nuclear fission*) or by fusing light atoms into heavier atoms (*nuclear fusion*). Because elements such as uranium and plutonium that are capable of sustaining fission reactions have extremely low cosmic abundances, nuclear fission cannot be responsible for the sun's energy generation. Nuclear fusion can, however. Hydrogen, the most abundant of the elements in the sun, can be transformed into helium by means of either the *proton-proton (PP) cycle,*

$$H^1 + H^1 \rightarrow H^2 + \text{positron} + \text{neutrino}$$
$$H^2 + H^1 \rightarrow He^3 + \text{energy (gamma ray)}$$
$$He^3 + He^3 \rightarrow He^4 + H^1 + H^1 + \text{energy}$$

or the *carbon (CN) cycle,*

$$C^{12} + H^1 \rightarrow N^{13} + \text{energy (gamma ray)}$$
$$N^{13} \rightarrow C^{13} + \text{positron} + \text{neutrino}$$
$$C^{13} + H^1 \rightarrow N^{14} + \text{energy}$$
$$N^{14} + H^1 \rightarrow O^{15} + \text{energy}$$
$$O^{15} \rightarrow N^{15} + \text{positron} + \text{neutrino}$$
$$N^{15} + H^1 \rightarrow O^{16} \rightarrow C^{12} + He^4$$

The superscripts denote the atomic weights of the elements involved. A *positron* is an electron having a positive electric charge and a *neutrino* is a mysterious particle in nuclear physics that has no mass or electric charge but is capable of carrying energy and can slip through matter like a nuclear phantom.

Theoretical calculations indicate that the temperature at the center of the sun is 10 to 20 million degrees Kelvin and the pressure is over 400 billion times that of the earth's atmosphere. Under such conditions fusion by both the proton-proton cycle and the carbon cycle can occur. Further investigations have shown that the proton-proton cycle accounts for 80 percent of the sun's radiant energy and the carbon cycle for the remainder. Because the temperature and pressure in the sun's interior become less extreme at greater distances from the center (for example, at a distance of $R_\odot/2$ from the center of the sun, the temperature drops to 3 million degrees Kelvin and the pressure to less than 1 billion atmospheres), it is assumed that virtually all of the fusion processes within the sun occur at or very near the center. The energy generated in the center slowly makes its way outward in the form of photons that are absorbed and reemitted by the highly ionized atoms in the sun's deep interior (see Fig. 8.1). At a distance of $0.8 R_\odot$ from the center, gigantic convection currents carry this radiant energy to the surface of the sun. The tops of these convection currents are observa-

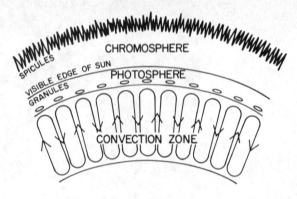

CORONA

CHROMOSPHERE

SPICULES

VISIBLE EDGE OF SUN

PHOTOSPHERE

GRANULES

CONVECTION ZONE

RADIATIVE ZONE

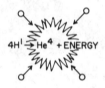

$4H^1 \rightarrow He^4 + ENERGY$

Fig. 8.1 A schematic diagram of the solar interior and atmosphere.

ble as the *solar granulation* and give the sun's outer layers their mottled or granulated appearance (see Fig. 8.2).

THE SOLAR ATMOSPHERE

The solar atmosphere consists of three regions: the photosphere, the chromosphere, and the corona.

The Photosphere. As the distance from the sun's center increases, the temperature and pressure of these interior layers decrease until the gaseous convective currents can no longer be maintained. The end of the convection zone roughly marks the boundary between the solar interior and the sun's lowest atmospheric layer, the photosphere. The light emanating from the photosphere (light sphere) defines the sun's visible disk. Its spectrum contains hundreds of absorption lines that reveal the presence of more than sixty elements and molecules in the photospheric gas. In addition to line absorption, the photosphere spec-

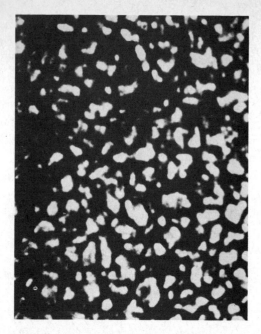

Fig. 8.2 The solar granulation. Each convective cell is about 800 km across. (NASA.)

trum shows continuous absorption, the source of which is the negative hydrogen ion, a hydrogen atom that for a very short time picks up a free electron and thus has a negative charge. As negative hydrogen ions form and dissociate, continuum radiation is absorbed and reemitted in a random fashion until it finally reaches the end of the photosphere and escapes into space.

The light coming from the center of the solar disk appears brighter than that coming from the edges (see Fig. 8.3). This phenomenon, referred to as *limb darkening,* arises out of the fact that the light observed at the center of the sun originates in the deeper, hotter regions of the photosphere and that observed at the edges originates in the higher, cooler regions. A detailed study of the limb-darkening effect has allowed astronomers to develop a reasonably accurate picture of the physical conditions present in the photosphere. The temperature at the top of the photosphere is found to be about 4500°K; it increases to 8000°K some 400 km into the sun's interior. The photospheric pressure increases over the same interval from about 10^{-2} atmosphere to just under 1 atmosphere.

The Chromosphere. Just above the photosphere is the chromosphere (color sphere), a layer about 16 thousand km thick whose den-

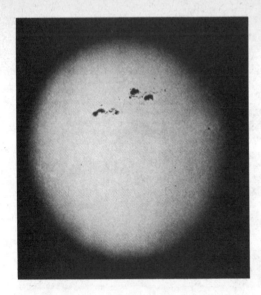

Fig. 8.3 Limb darkening of the solar disk. Note the rapidly decreasing brightness of the sun's disk near the edge of the sun. (Courtesy of the Hale Observatories.)

sity decreases with increasing altitude but whose temperature increases dramatically over the same interval from 4500°K to over 100,000°K. For many years observations of the chromosphere could be made only in the few seconds preceding a total eclipse of the sun. At this time the photosphere is covered, and a reddish crescent of light from the chromosphere can be briefly observed. The spectrum of this light, called a *flash spectrum,* exhibits elemental abundances much the same as those found in the photosphere. Because of the higher temperatures, however, atoms in the chromosphere are in much higher states of ionization and excitation and thus display slightly different sets of spectral lines. It was in the chromospheric spectrum that the "element of the sun," helium, was first discovered, years before it was isolated and identified.

The difficulty of observing the chromosphere has been overcome by the invention of the *coronagraph,* a device in which the sun is eclipsed by optical means, thus permitting a more leisurely study of this region than is possible during an eclipse. The chromosphere can also be studied by use of a *spectroheliograph,* which allows the sun to be photographed in the light of a very narrow spectral line such as the hydrogen alpha line at 6563 Å.

In the upper region of the chromosphere can be seen a series of bright jets called *spicules* (see Fig. 8.4). These jets shoot through the

Fig. 8.4 Spicules in the solar chromosphere. (Sacramento Peak Observatory, Association of Universities for Research in Astronomy, Inc.)

chromosphere to heights of several thousand kilometers above the photosphere at velocities of 15 to 25 km/sec. Although each spicule lasts only from 2 to 5 minutes, about 100,000 of these features can be seen evenly distributed around the solar limb at one time, giving the appearance of a brush fire. It is thought that the spicules are involved in the transport of energy through the chromosphere and as such are closely related to the solar granulation in the lower photosphere.

It is also in the chromospheric layer of the solar atmosphere that protons and electrons as well as trace amounts of heavier nuclei are "boiled off" the sun and thrust into the interplanetary medium. As these particles leave the sun they are thermally accelerated by the high coronal temperatures and eventually reach speeds of several hundred kilometers per second. This continual ejection of high-velocity atomic particles into the interplanetary medium is commonly referred to as the *solar wind*. It has been measured extensively by artificial satellites as well as by deep-space probes.

The Corona. During the total phase of a solar eclipse, the sun is surrounded by a pale glow of light called the corona (crown). The corona is the outermost region of the solar atmosphere and extends thousands of kilometers into space. It often displays a complicated structure of streamers, which suggests the presence of strong localized magnetic fields (see Fig. 8.5).

The outer portion of the corona, or *F corona*, exhibits a reflected-absorption line spectrum and is thought to be due to an inner extension of the interplanetary dust that gives rise to the zodiacal light (see Chapter 7) at greater distances from the sun. The inner portion of the corona, or *K corona*, exhibits a spectrum consisting of a number of emission lines superimposed on a weak continuum. For many years these lines defied identification and were thought to arise from yet another undiscovered element, "coronium." However, the "coronium" lines were shown to arise from atoms of iron, nickel, and calcium from which 13 or more electrons have been stripped. The

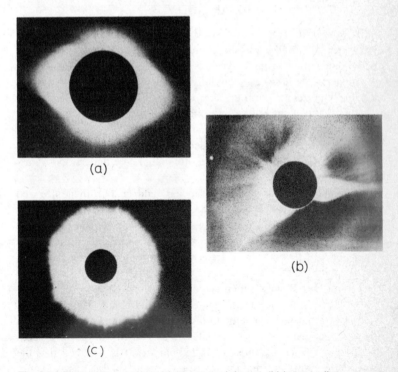

(a)

(b)

(c)

Fig. 8.5 The solar corona at (a) sunspot minimum, (b) intermediate sunspot numbers, and (c) sunspot maximum. [(a) High Altitude Observatory, National Center for Atmospheric Research, sponsored by the National Science Foundation; (b), (c) Yerkes Observatory photographs.]

coronal temperature required for such a high degree of ionization is roughly 2 million degrees Kelvin, a value that is confirmed by analyses of the profiles of the emission lines. The density of the corona, however, is little more than that of the interplanetary medium, a density so low that despite the high coronal temperature the total thermal energy emitted by the corona is small compared to that which arises from the lower-temperature, but much higher-density photosphere. The means by which the corona is heated to such a high temperature is thought to be related to the deposition of large amounts of energy from the denser convective layers of the sun at the base of the corona by the granulation and the spicules.

SOLAR ACTIVITY

Sunspots have been known since the time of Galileo and possibly even earlier. In 1838 it was discovered that the number of spots visible on the sun varies in an 11-year cycle. More detailed investigations have revealed a wealth of solar activity whose frequency and intensity are closely correlated with the *sunspot cycle*. Thus, astronomers often refer to the sun as the *quiet sun* at times of sunspot minima and the *active sun* at times of sunspot maxima. The physical processes by which the various aspects of the solar activity cycle occur and their relationship to one another are, to date, not completely understood by astrophysicists.

Sunspots. Sunspots appear, disappear, and change their size and shape as they advance across the sun's disk. Sunspots are about 1500°K cooler than the surrounding solar gas and hence appear darker. The lower temperature of sunspots is thought to be due to the reduction in energy flow to these areas by the strong magnetic fields associated with them.

Faculae and Plages. Often a region develops on the solar surface that has the appearance of a bright, granulated cloud. These regions, called faculae, usually develop in the vicinity of sunspots or in regions with stronger than average magnetic fields. When viewed in a spectroheliograph, such features are called *plages*. It is assumed that the faculae and plages are regions where the chromospheric gas is changing its state of ionization and/or excitation, thereby emitting photon radiation.

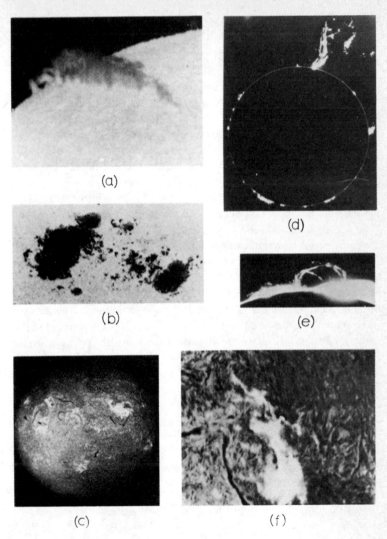

Fig. 8.6 Some aspects of the solar activity: (a) a filament, (b) sunspots, (c) faculae (light areas) and filaments (dark areas), (d) a limb flare and prominences, (e) a loop prominence, and (f) a flare. [(a), (e) Sacramento Peak Observatory, Association of Universities for Research in Astronomy, Inc.; (b), (f) courtesy of the Hale Observatories; (c) photograph by Lockheed Solar Observatory; (d) photograph taken at the University of Hawaii's Mees Solar Observatory, Institute for Astronomy, Haleakala, Maui.]

Prominences and Filaments. Periodically huge jets of gas will, in a matter of minutes or hours, shoot thousands of kilometers into space from the edge of the sun's surface (see Fig. 8.6). These jets, called prominences, are often twisted and bent by magnetic fields and take on different shapes. Prominences are usually observed at times of sunspot maxima, but, like other aspects of solar activity, they sometimes occur at times of sunspot minima. Prominences that are not at the solar edge project onto the solar disk as dark, wispy features called filaments.

Flares. Flares are the most impressive manifestations of solar activity. Sometimes a highly localized area of the sun suddenly brightens up over a period of a few minutes. Associated with this event are a wide variety of emissions, including X rays, radio bursts, and high-velocity atomic particles. Such outbursts can have a considerable impact on the earth and its atmosphere, and among other effects, can produce partial communications blackouts and aurorae.

Other Changes. A number of other changes in the solar atmosphere are related to the solar activity cycle. At sunspot maximum, the corona is almost completely symmetrical about the solar disk and extends out at a considerable angular distance; at sunspot minimum, it takes on a more flattened appearance and exhibits weak polar rays (see Fig. 8.5). The variation in the strength of certain lines in the spectrum of the solar corona is in phase with the solar activity cycle.

Most interesting of all, perhaps, is the fact that certain changes in the earth's environment seem to be related to the solar activity cycle. For example, botanists have found that the spacing in tree rings, indicative of tree growth rates and hence climatic conditions, can be correlated with the solar activity cycle. Because there are so many other variables, however, such potentially important correlations between the solar cycle and the earth's environmental conditions are unfortunately of a very indirect nature.

REVIEW QUESTIONS

1. Find the mean density of the sun using measured values for the solar mass and radius. What does the result suggest about the nature of the sun? *Ans.:* 1.4 g/cm^3.
2. Suppose that you live on a planet where the solar constant is observed to be 10 ergs/sec/cm^2. If the distance between the planet and the local sun is 2×10^{14} cm, find the total luminosity of the local sun. *Ans.:* 5.0×10^{30} ergs/sec.

3. Calculate the lifetime of the sun if it were to convert all of its mass into nuclear energy at its present rate of energy production. *Ans.:* 10^{11} years.

4. Describe the interior of the sun.

5. Compare and contrast the photosphere, chromosphere, and corona.

6. Describe the various observable aspects of the ''active'' sun.

9

Properties of Stars

Since the time of Bruno in the late sixteenth century, astronomers have suspected that stars are bodies similiar to our sun, but so far from the earth that they are seen as mere dots of light in the night sky. The successful measurement of stellar parallaxes in 1838 confirmed this assumption and provided the first insight into the nature of these objects. Astronomers have since discovered that although the stars are indeed similiar to our sun, a great many possess qualities that are vastly different.

INDIVIDUAL STARS

Many of the properties of stars can be determined by measurements of individual stars. Other properties, however, can be determined only by the analysis of binary star systems.

Stellar Distances. The direct method by which stellar distances are obtained is the trigonometric measurement of stellar parallax (see Fig. 3.4). Photographic plates are taken of a star when it is at right angles to the east of the sun and 6 months later when it is at right angles to the west of the sun. If the star is sufficiently close to the sun, it will exhibit a small shift or parallax relative to the background stars. By knowing the size of the earth's orbit and the size of the parallax, the distance to the star can be obtained by using the triangulation techniques described in Chapter 3. In most cases a stellar distance can be reliably measured up to about 50 parsecs using the method of trigonometric parallax.

If the mean absolute magnitude $\overline{M}$ of a certain class of stars is known and if one observes the apparent magnitude m of a star belonging to that class, the distance r to the star in parsecs can be obtained by using the distance modulus formula $m - \overline{M} = 5 \log r - 5$ (see Chapter 3), where the absolute magnitude of the star of interest is assumed to be the mean for the class of stars to which it belongs. Such a determination of distance is referred to as a *spectroscopic parallax*.

Stellar Motions. Stars display in varying degrees two separate types of motion, proper motion and radial velocity. *Proper motion* is the angular motion μ observed for a star per unit time. Although some stars such as Barnard's star exhibit proper motions over 1 arcsecond per year, most proper motions are either small fractions of a second of arc per year or are so small that they cannot be detected at all. The *radial velocity* (V_r) of a star is the component of a star's velocity that is directed along the observer's line of sight. Radial velocities are determined spectroscopically using the Doppler effect. If the distance to the star is known, its component of velocity T perpendicular to the line of sight, called the *tangential velocity*, can be calculated from the relation $T = 4.74 \, \mu r$ (km/sec), where μ is in arcseconds per year and r is the radius. If both the radial velocity and the tangential velocity of a star are known, the vector sum (see Fig. 9.1) of these velocities yields the

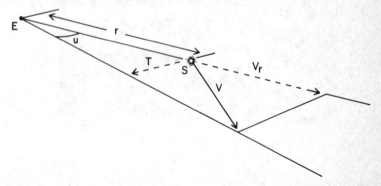

Fig. 9.1 The relationship between a star's radial velocity V_r its tangential velocity T, and its space velocity V. The distance to the star is r parasecs and its proper motion is μ arcseconds.

space velocity, V, for the star, which is a true measure of the direction and speed of the star's motion relative to the sun.

Stellar Temperatures. Temperatures of stars can be determined by a wide variety of methods. One simple technique is to assume that the energy output of the star can be approximated by a blackbody curve (see Chapter 3). The *effective temperature* of the star is then defined as the temperature that a blackbody would have if it produced the same

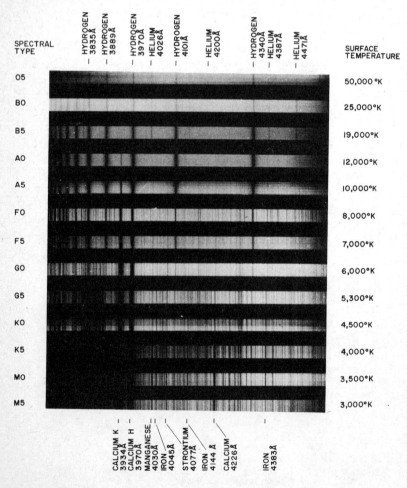

Fig. 9.2 The sequence of spectral types along with their respective surface temperatures. Some prominent absorption lines are also identified. (The Kitt Peak National Observatory.)

total energy per unit area as the star. The *Wien's law temperature* is the temperature that a blackbody would have if its wavelength of maximum light were equal to that of the star, and the *color temperature* of a star is the temperature that a blackbody would have if its photometric B-V color were the same as that of the star, assuming that no interstellar reddening is present.

The above temperatures are based on the star's continuum radiation. Temperatures can also be determined from the nature of the absorption lines present in the star's spectrum. The temperature of a star dictates which energy levels in an atom will be occupied by the electrons and hence which upward transitions they can make. Thus stars having essentially the same composition will exhibit a wide variety of absorption-line spectra, depending on the temperature of the star (see Fig. 9.2). These large differences allow astronomers to estimate the temperatures of stars to within a few percent by simply noting the appearance of the spectrum. To facilitate the study of a star's spectrum, astronomers have divided stars into the *spectral types* O, B, A, F, G, K, and M. These types are based solely on which absorption lines are present in a star's spectrum and how strong these lines appear to be, and serve as useful indicators of stellar temperatures. The temperature ranges and spectral characteristics of these spectral classes are summarized in Table 9.1. To take into account objects whose spectra are intermediate between two spectral classes, each class is subdivided by assigning to it a number between 0 and 9. For example, an A5 star is halfway between an A star and an F star in its spectral characteristics. The sun is a G2 star and, as such, possesses more of the qualities of a G star than those of a K star.

Table 9.1.   Stellar Spectral Classes

Class	Approximate Temperature Range (°K)	Outstanding Spectral Characteristic(s)
O	>25,000	Ionized helium lines
B	12,000–25,000	Neutral helium lines
A	8,000–12,000	Hydrogen lines
F	6,000–8,000	Ionized metallic lines
G	4,500–6,000	Neutral metallic lines
K	3,500–4,500	Neutral metallic lines and weak TiO molecular bands
M	<3,500	TiO molecular bands

Stellar Luminosities. The *luminosity* of a star is the total energy output of that star at all wavelengths per unit time. It was noted some years ago that a set of stars of the same spectral type often had a wide range of luminosities. Moreover, several luminosity classes could be identified from their effects on certain lines in the spectra of stars, namely:

Ia   Most luminous supergiants
Ib   Less luminous supergiants
II   Bright giants
III   Normal giants
IV   Subgiants
V   Dwarfs (main sequence stars)

It should be noted that these luminosity classes apply only within the framework of a given spectral type. For example, an M0 III star is certainly more luminous than an M0 V star but less luminous than an O8 V star. A complete spectral classification of a star thus includes its spectral type as well as its luminosity class. The complete spectral designation of the sun in this classification system, called the *Morgan-Keenan* or *MK system,* would be G2 V.

The actual energy output of each luminosity class within a given spectral type can be obtained by first calculating the star's absolute magnitude from the distance modulus formula (see Chapter 3), assuming that the distance is known, and then employing the relation

$$M_{bol*} - M_{bol\odot} = -2.5 \log\left(\frac{L_*}{L_\odot}\right)$$

in which the symbols * and $\odot$ stand for star and sun, respectively. Because the sun's absolute bolometric magnitude and its luminosity is known, the star's luminosity can be obtained.

The fact that certain spectral lines are sensitive to a star's luminosity in a quantitative way provides another method of determining the energy output of a star. The luminosity effects on the emission cores of the H and K lines of calcium in the violet region of the spectrum for several G and K stars are illustrated in Fig. 9.3. Stellar luminosities obtained from this type of calculation are found to range from 10^6 to 10^{-4} that of the sun.

Stellar Radii. Despite the fact that stars are very large objects whose angular diameters often span millions of kilometers in space, their even greater distances from the earth preclude any direct detec-

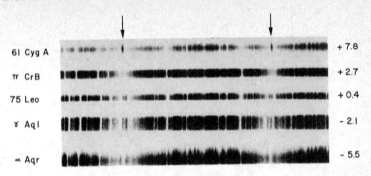

Fig. 9.3 Luminosity effects in stellar spectra. Note the increasing width of the emission cores of the calcium lines as luminosity increases. (Courtesy of the Hale Observatories.)

tion of a stellar disk. Astronomers have thus devised several indirect methods by which stellar radii and diameters can be measured. If the effective temperature and luminosity of a star are known, then the star's radius can be determined from the relation

$$L = 4\pi R^2 \sigma T_e^{\,4}$$

provided that the star's radiation can be approximated by a blackbody radiation curve.

Stellar radii can also be obtained by directly measuring the angular diameter of the star through the use of a device called an interferometer (see Fig. 9.4). As noted in Chapter 3, because of the phenomenon of diffraction, a single-point source of light is imaged by a circular aperture as a finite disk of light surrounded by a series of concentric circular fringes, and a pair of point sources of light will be imaged as a pair of finite light disks surrounded by overlapping fringe patterns. A stellar image can be regarded as arising from two point sources, each of which represents half of the star's radiating disk. By feeding the radiation from the star into a telescope or electronic analyzer through two movable reflectors, the double fringe pattern can be produced for the star. At the proper separation of the reflectors, the fringes can be made to overlap in such a way that the resultant superposition of fringe patterns is a uniformly bright image. The angular diameter α of the object is then equal to a constant times the value (λ/s_0) where λ is the wavelength of the incoming radiation and s_0 is the separation of the reflectors required for the disappearance of the

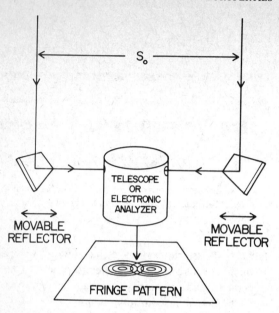

Fig. 9.4 The stellar interferometer. By adjusting the distance s_0 so that the fringe pattern is uniformly bright, a determination of the star's angular diameter can be made if the wavelength of the incoming light and the distribution of light across the star's disk are known (or assumed).

fringes. The value of the constant depends on the assumptions made concerning the distribution of brightness across the stellar disk as well as the units employed for α, λ, and s_0. Once the angular diameter of the star is measured, the linear diameter and radius are easily obtained if the distance to the star is known.

Magnetic Fields. If the atoms of a gas are placed in a magnetic field, certain absorption lines appear to split into two or more components. This effect in spectroscopy, called the *Zeeman effect,* arises from the fact that the magnetic field causes a splitting of some of the atom's energy levels. If the magnetic field is strong enough or if the resolution of the spectrograph is high enough, each of the individual component lines can be resolved. For most astronomical objects, however, the individual components cannot be separated and the spectral line appears only to be broadened. By studying the degree of this magnetic broadening, astronomers can estimate the strengths of stellar magnetic fields. The peculiar star 53 Cam, for example, has a magnetic field over 10^5 times that of the earth.

Turbulence. In analyzing the spectra of the cooler stars, especially the cool supergiants, it was found that the spectral lines present were much stronger than would be expected for the given star's effective temperature. Further investigation of this phenomenon revealed that the strengths of the lines are enhanced by a Doppler broadening arising from large-scale vertical convective motions or gas in the atmospheres of these stars. These convective motions are called turbulent motions or turbulence. Detailed analyses of turbulent motions indicate that they can reach velocities as high as 8 km/sec.

BINARY STARS

In 1650 the Italian astronomer Jean Baptiste Riccioli found that the star Zeta Ursae Majoris (Minor) was in reality a pair of stars so close together in the sky that they could not be seen as separate bodies except through a telescope. Other accidental discoveries of these so-called double stars continued until the last quarter of the eighteenth century, when Sir William Herschel of England began a systematic search for them. Herschel sought these objects for their possible use in determining the long-elusive stellar parallax. He found that there were far too many double stars in the sky to be attributed simply to a chance optical alignment of two stars at different distances along a given line of sight. In 1803 Herschel demonstrated conclusively the existence of binary star systems, that is, systems in which two stars are gravitationally associated and revolve about one another.

Types of Binaries. Binary stars are classified according to the method by which they can be detected. A given binary pair might fit more than one category. The three types are visual binaries, eclipsing binaries, and spectroscopic binaries.

VISUAL BINARIES. Visual binaries are binary stars in which both members are visible. Their mutual revolution can be detected over several years by systematically measuring the position angle and separation for the two stars as shown in Fig. 9.5. Some star pairs that are actually at great distances from one another appear to be binaries because they lie along the line of sight when viewed from the earth. Such pairs, called *optical doubles*, can usually be recognized by the fact that their proper motions, parallaxes, and radial velocities are significantly different.

ECLIPSING BINARIES. Eclipsing binaries are binary systems with an orbital plane that passes through our line of sight. Although they are not resolvable, their presence is made known when one of the stars

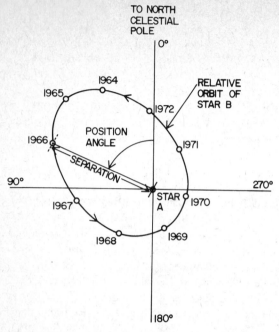

Fig. 9.5 A hypothetical visual binary showing the position angle and separation of star B relative to star A.

passes in front of the other and eclipses it. The resulting drop in the total brightness of the system is readily detected by photometric means.

SPECTROSCOPIC BINARIES. Other unresolvable binaries are detected by Doppler shifts in their spectral lines that vary with time. These variations are thought to be due to actual changes in the radial velocity component of one of the stars' motion as it orbits its companion. Such systems are called spectroscopic binaries. If the absorption lines from only one of the stars are detectable, the system is referred to as a single-line spectroscopic binary; if lines can be observed from both components of the system, it is called a double-line spectroscopic binary.

Multiple Star Systems. A few star systems exist that contain more than two stars but not a sufficient number to be categorized as star clusters. In some multiple star systems, such as Alpha Centauri and Zeta Cancri, a close visual binary system is orbited by a third, more distant star. In others, each component of a visual binary is itself a close spectroscopic binary system, thus making a quadruple star system. The Xi Ursa Majoris system is an example of this type of arrangement. The amazing Castor system in Gemini has an Alpha Centauri-type arrangement of three stars, but each of these stars is in turn a spectroscopic binary. Thus the Castor system consists of six stars.

Each of these multiple star systems behaves as if it were a set of two-body systems just as the earth-moon-sun system is in effect a mutually revolving twin-planet entity itself revolving about the sun. The analysis of the physical properties of multiple star systems is thus similar to that employed for binary stars.

Information Provided by Binary Stars. The analysis of binary stars provides data that can be used to determine the masses and dimensions of many other stars. The study of binary systems also gives us evidence for stellar rotation.

STELLAR MASSES. Aside from the sun, binary stars provide astronomers with the only source of data for determining stellar masses because these are the only systems in which an orbiting body is bright enough for its behavior to be observed as it moves in its companion's gravitational field. From such data came the mass-luminosity relationship.

Determining the Mass of a Visual Binary. If we know the distance of a visual binary from the earth and we measure the angular separation between the two stars, we can obtain the mean distance a between the stars (in seconds of arc). The sidereal period P of the revolution (in years) is obtained by observation over a long time of the stars' revolution. Kepler's harmonic law (restated by Newton; see p. 62),

$$\frac{a^3}{P^2} = \boldsymbol{M}_1 + \boldsymbol{M}_2$$

then yields the sum of the masses in solar units. Only the *sum* of the masses can be determined from this formula. Often, however, careful studies of the binary's radial velocity and proper motion reveal small oscillating motions of each star about the system's center of mass, or *barycenter*. The two stars' motions about the barycenter are such that the product of the mass and the distance to the barycenter is the same for both stars. By measuring the relative amplitudes of these oscillating motions, astronomers can determine the mass ratio $\boldsymbol{M}_1/\boldsymbol{M}_2$ of the system and thus the individual masses of the two stars.

Determining the Mass of a Spectroscopic Binary. If the measured

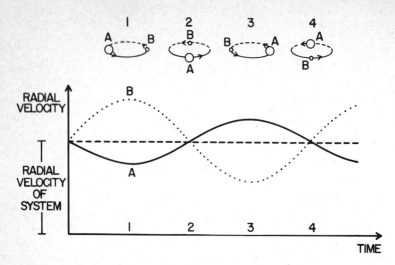

Fig. 9.6 A diagram of a spectroscopic binary system. As each star orbits the other, the variable radial velocity curves are produced.

radial velocity for a spectroscopic binary is plotted against time, the resulting curve (*radial velocity curve*) indicates the behavior of the radial component of the star's motion with respect to the binary's center of mass (see Fig. 9.6). If the system is a single-line system, the astronomer can obtain from the velocity curve only the value of the *mass function*,

$$f(M_1) = \frac{(M_1 \sin i)^3}{(M_1 + M_2)^2}$$

where i is the inclination of the binary's orbital plane to a plane tangent to the celestial sphere. For a double-line system, two radial velocity curves may be plotted and a mass ratio M_1/M_2 determined. Only if the system is double-lined *and* eclipsing can the individual masses be determined for a spectroscopic binary. In such a system, the inclination is a right angle and the two values determined for the mass functions from the radial velocity curves yield a set of two equations involving two unknown masses, which can be readily solved by algebraic means.

Mass-Luminosity Relationship. If stellar mass is plotted against the corresponding luminosity for each of the binary systems having well-determined individual masses, most of the stars fall along the sequence of points shown in Fig. 9.7. This relationship between a star's mass and its intrinsic energy output is called the mass-luminosity law. The mass-luminosity law states that the luminosity or total energy output of a star is roughly proportional to the star's mass raised to the power 3.5. Using this relationship, astronomers can estimate the mass of any star whose luminosity can be determined. Also, any model stars (see p. 152) that are constructed in a computer must obey this relationship, and as such, the mass-luminosity law provides an excellent check on the accuracy of the model stars that astronomers have constructed by theoretical means.

STELLAR DIAMETERS. Eclipsing binaries can provide useful information on the dimensions of stars if the light curve (light plotted against time) associated with the eclipse is carefully observed. The process is illustrated for an idealized eclipsing binary in Fig. 9.8. As the eclipse proceeds, an ever-increasing amount of light from the star being eclipsed is blotted out. Finally, the eclipsing star is completely

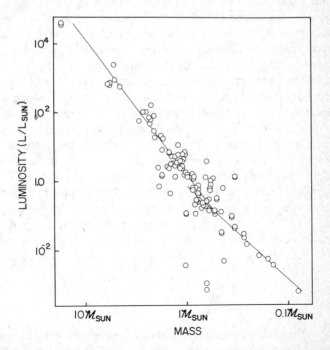

Fig. 9.7 The mass-luminosity relation for binary systems whose masses, distances, and brightnesses are well determined.

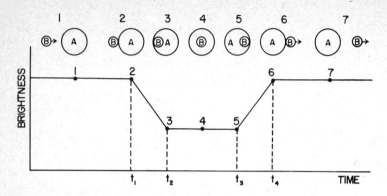

Fig. 9.8 An idealized eclipsing binary and its corresponding light curve.

within the disk of the eclipsed star and can blot out no more of the latter star's light. The light curve then displays a flat bottom until the eclipsing star begins to move from its position in front of its companion. As the eclipse is ending, the brightness gradually reaches its former level. If the velocities of the stars can be determined spectroscopically, then the diameter of each is obtainable by multiplying the velocity of the eclipsing star by appropriate time intervals which are read off the light curve (Fig. 9.8). This is, of course, a highly idealized situation. In practice, difficulties arise because, among other things, the eclipses are usually not central (i.e., the center of one star does not pass over the center of the other), the stars are often close together and thus tidally distorted, and the stars are probably not uniformly illuminated but show limb darkening similar to, but not the same as, that observed for the sun.

DYNAMIC PARALLAX. In some instances, it is possible to obtain the distance to a binary system by observing its motion and making assumptions as to its mass. In this method, an initial guess is made for the two masses of the binary, usually one solar mass each, and a value for the distance r is calculated using the following version of Kepler's harmonic law:

$$M_1 + M_2 = \frac{(r \times a'')^3}{P^2}$$

where a'' is the measured mean distance between the components of the binary in arcseconds. The distance modulus formula is then used in conjunction with the measured apparent magnitudes for the system to calculate the luminosities of each component, and a new mass estimate is obtained for each star from the mass-luminosity law. The process reverts back to a recalculation of r based on the new mass estimates and the above process is repeated. The iteration, or repetition of calculation, is continued until two successive distance determinations agree to within some specified amount.

STELLAR ROTATION. It has long been recognized that the sun rotates on its axis at a rate of about once every 27 days, but it was not until the early decades of this century that astronomers were able to assemble evidence that the distant stars also rotate, and often at enormous velocities. The most direct evidence for stellar rotation is found among the eclipsing binary stars. For some of the eclipsing systems it has been noted that the radial velocity curve of the eclipsed star will ex-

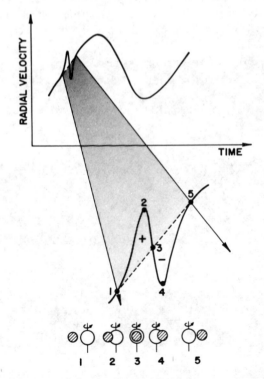

Fig. 9.9 The Rossiter effect in an eclipsing binary system. As the rotating primary star is eclipsed, a single "wobble" in that star's radial velocity curve is produced.

hibit an anomalous single oscillation from the mean curve at the exact time of the eclipse. This behavior is referred to as the *Rossiter effect* and is explained by the fact that as the eclipsing star covers up first one half of the rotating star and then the other, the observer sees in turn that half of the star that is rotating toward (or away from) him and then that half rotating away from (or toward) him. The observed result is a single Doppler oscillation superimposed on the normal radial velocity curve as shown in Fig. 9.9.

Larger rotation rates also tend to impart a "washed-out" appearance to spectral lines, which become larger with increasing rates of rotation (see Fig. 9.10). Because the angular inclination i of the axis of rotation of the star relative to the observer cannot be determined for the distant stars, one can only determine the value of $V_{rot} \sin i$, or the component of the rotational velocity V_{rot} that is directly toward or away from the observer. The most rapidly rotating stars are the B and early A (A0–A5) stars, some of which must be on the verge of rota-

Fig. 9.10 The effect of rotation on stellar absorption lines. The lines in the lower spectrogram have a "washed-out" appearance due to that star's rotation, whereas the spectrogram above is from a star whose rotation rate is much less. (The Kitt Peak National Observatory.)

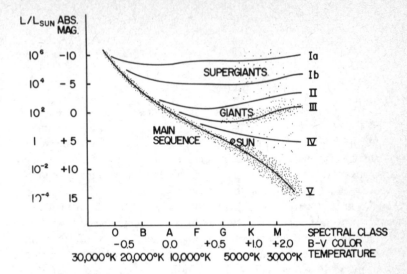

Fig. 9.11 The H-R diagram for stars in the solar neighborhood.

tional disruption because of their extremely high rates of rotation. Stars cooler than G0, such as the sun, are observed to have considerably reduced rotational velocities.

THE HERTZSPRUNG-RUSSELL DIAGRAM

If a plot is made of surface temperature versus absolute magnitude for stars for which these quantities have been accurately measured, the resulting distribution of points takes on the appearance shown in Fig. 9.11. Such a plot is referred to as a Hertzsprung-Russell or H-R diagram after its co-discoverers, E. Hertzsprung of Denmark and H. Russell of the United States. It is also known as a spectrum-luminosity diagram.

The areas of the H-R diagram where the largest number of stars are found are the *main sequence* and the *giant branch*. The sun occupies a position on the main sequence below and slightly to the left of the giant branch. Because these points on the H-R diagram are experimentally determined over such a short time period relative to the lifetimes of stars, they represent a "frozen" picture of stars in their various stages of evolution, much as a photograph taken of a garden or forest is a frozen picture of the state of the plants present at that time.

For this reason, the study of the various facets of the H-R diagram is of considerable importance in understanding how stars evolve during their life cycles.

MODEL STARS

To help understand the physical processes at work in the interiors and the atmospheres of stars, astronomers often construct complete mathematical descriptions of stars called stellar models or model stars.

In constructing a theoretical model interior for a star, the astronomer assumes for the star a given mass and composition. A number of physical relationships, such as the equation of state for a gas, that are believed to govern the workings of the star are then used to generate a system of mathematical equations which in turn are solved to obtain a complete physical description of the star, including its effective temperature, luminosity, and radius, as well as the temperature, pressure, density, and energy output at any point in the star's interior.

The model interiors developed for the ordinary stars are, for the most part, confirmed by direct determinations of the stars' physical properties. For example, the straight-line characteristics of the mass-luminosity law and the main sequence on the H-R diagram can be theoretically reproduced from model star calculations. Model interiors also predict that the basic structure of the interiors of cool stars is that of a radiative core region surrounded by a convective envelope of gases, whereas the structure of the hotter stars is the reverse. This prediction is upheld in part by the existence of solar granules, which are known to be the tops of convective currents present in the envelope layers of the solar interior.

Of particular interest is the use of stellar models to investigate in a few minutes of computer time what happens to a star's physical properties as a result of millions of years of radiating as a star. Such studies strongly suggest that the stars are not static entities, but are in fact evolving and changing over very long periods of time (see Chapter 12).

A complete mathematical description of the various layers of the atmosphere of a star can be compiled using techniques similar to those employed in the development of model interiors. Such descriptions are called *model atmospheres* and are of great use in the studies of such observational properties of stellar atmospheres as element abundances, pressures, and so on.

REVIEW QUESTIONS

1. A certain star exhibits a parallax of 0.050 arcsecond. What is the distance to the star? *Ans.: 20 pc.*

2. A star exhibits a spectrum similar to that of the sun and has an apparent magnitude of +10. What is the distance to this star? *Ans.: 100 pc.*

3. Describe the various types of stellar motions.

4. A star has an absolute bolometric magnitude of 0.0. What is the star's luminosity in ergs/sec? Assume that $M_{bol\odot} = 5.0$ and $L_\odot = 4 \times 10^{33}$ ergs/sec. *Ans.: 4×10^{35} ergs/sec.*

5. Describe the spectrum of an F5 star.

6. How are stellar temperatures determined?

7. Calculate the radius of a star having a temperature of 6000°K and a luminosity of 4×10^{33} ergs/sec. *Ans.: 7×10^{10} cm.*

8. Star A "wobbles" about its barycenter with star B at a distance that is twice as large as that of star B. The period of revolution for the system is 3 years, and the separation between star A and star B is 3 AU. What are the individual masses of stars A and B? *Ans.: Star A = $1M_\odot$, Star B = $2M_\odot$.*

9. Explain how stellar radii can be determined from observations of eclipsing binary systems.

10. Compare and contrast dynamic parallax, trigonometric parallax, and spectroscopic parallax.

11. How are binary stars detected?

10

Peculiar Stars

In addition to the typical stars, which are described in Chapter 9, there exist a few thousand stars whose properties are unusual. Such objects are believed to represent various stages in the lives of stars and thus can provide a better understanding of stellar evolution.

STARS HAVING UNUSUAL ABUNDANCES OF ELEMENTS

Spectroscopic analysis of stellar atmospheres shows that stars are composed primarily of hydrogen (about 90 percent) and helium (about 10 percent). The relative abundances of the elements heavier than helium are roughly the same as those found in solar cosmic rays, the earth's crust, and meteorites. A few stars, however, exhibit significant variations from the typical relative abundances, and because it is not likely that these abundance anomalies were present in the material from which the star was formed, they are assumed to have arisen from nuclear processes occurring during the star's lifetime. Thus, a study of these abundance anomalies and the nuclear processes that could give rise to them yields important clues as to the nature of the deep interior of a star, especially a star in an advanced stage of its life cycle.

Helium Stars. A few stars exhibit most of the spectral characteristics of B stars but have lines of hydrogen that are weak or completely absent; these stars also exhibit an enhancement of the lines of helium. It may be that the hydrogen-rich outer layers of the star have been blown off, thus baring the helium-rich inner layers. Another possibility is that the two regions have become thoroughly mixed by convection and the relative hydrogen abundance has decreased in the outer layers while the relative abundance of helium has increased.

Peculiar A Stars. Perhaps the most puzzling of all the objects with abundance anomalies are the peculiar A stars. These objects have spectra resembling those of normal A stars, but also display very strong lines of the more common metals such as silicon, chromium, and strontium and of the rare earth elements such as europium; moreover, the strengths of these lines vary from time to time. Peculiar A stars may also exhibit strongly variable magnetic fields, some of which change their magnitudes by thousands of gauss in a matter of a few days.

Metallic-Line Stars. Positioned on the H-R diagram near the main sequence and the F stars are the metallic-line stars. These objects exhibit the same continuum distributions and gross spectral characteristics as the A and F stars but contain metallic lines characteristic of somewhat cooler stars. These objects all seem to possess large amounts of turbulence and all are spectroscopic binaries.

Barium Stars. In the region of the late G and early K giants and supergiants are the barium stars. Not only are the barium lines enhanced for these objects, but also the lines arising from other heavy metals such as zirconium and lanthanum, thus suggesting an overabundance of these substances relative to the light metals such as vanadium, titanium, and iron (see Fig. 10.1).

Carbon Stars. In the normal cool red giants, the molecular bands of the light metal oxides, in particular titanium oxide and vanadium oxide, dominate the spectrum. Carbon stars are a class of cool stars having luminosities similar to those of the normal red giants but whose spectra are characterized by bands arising from molecules containing carbon, especially C_2, CH, and CN.

S Stars. The S stars, like the carbon stars, are found in the region of the red giants on the H-R diagram. Their spectra, however, are dominated by the molecular bands of heavy metal oxides such as zirconium oxide and lanthanum oxide. Quite often the carbon stars and S stars are also intrinsic variables having light variations similar to those exhibited by the Mira variables and the irregular variables, to be discussed below.

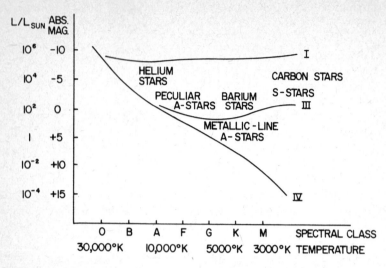

Fig. 10.1 The H-R diagram for stars having abundance anomalies.

INTRINSIC VARIABLE STARS

Binary stars that vary in brightness because they eclipse each other are sometimes called eclipsing variables. Intrinsic variable stars are those whose brightness varies as a result of changes arising from their physical properties. About 13,000 such stars are known at present. There are two main types of intrinsically variable stars: pulsating variables and eruptive variables.

Pulsating Variables. Nearly 10,000 intrinsic variables vary in brightness as a result of pulsations that occur in their atmospheric layers. The existence of such pulsations can be verified by observing the correlation between the measured brightness of the variable and its observed radial velocity (see Fig. 10.2).

Cepheid Variables. The cepheid variables, named for their prototype, Delta Cephei, are F and G supergiants that have nearly constant periods of pulsation ranging from 1 day to more than 50 days. During these pulsations, the cepheid rises rapidly to its maximum brightness and more slowly dims to its minimum light. Typically, the radius is altered by about 10 percent during this cycle with corresponding changes in the luminosity and surface temperature.

The cepheids can be divided into two categories: type I, or classical, cepheids, which are the brighter cepheids and are found in the regions of high gas and dust content along the Milky Way plane; and type II cepheids, or W Virginis stars, which are fainter and have a roughly spherical distribution around the nucleus of the Milky Way.

An important property of the cepheid variation is that the period P of the pulsation (light variation) and the mean absolute magnitude M (average luminosity) are given by the relation

$$M = A + B \log P$$

where A and B are measured constants. This relation is known as the period-luminosity law.

Because of their large intrinsic brightnesses, distinctive light variation, and adherence to a well-defined law, cepheid variables are used extensively as distance indicators. Such stars in galaxies and star clusters are readily identified from the characteristic nature of their light variation. The period of the light variation is then measured and the corresponding absolute magnitude calculated from the period-luminosity law. The absolute magnitude and the apparent magnitude m of the object, which is easily determined, provide enough information

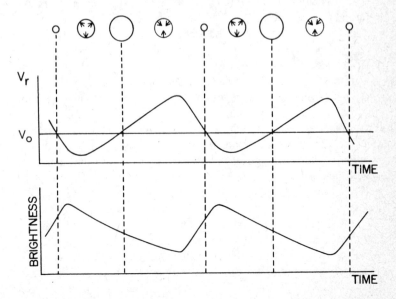

Fig. 10.2 The correlation among a pulsating variable's size (top line), radial velocity (middle graph), and apparent brightness (bottom graph).

to compute the distance modulus $m - M$ and hence the cepheid's distance. Care must be taken to identify correctly type I and type II cepheids, since the one to two magnitude difference in their absolute magnitudes can lead to a distance determination that is off by as much as a factor of 2.

RR LYRAE STARS. Below the cepheids in the H-R diagram and similar to them in properties are the RR Lyrae stars. However, their periods of light variation are much shorter, being in the range of 0.1 to 1.0 day. The absolute magnitude of RR Lyrae stars does not exhibit the period-luminosity relation as does that of the cepheids, but their mean absolute magnitude of 0.0 is roughly what would be expected for a type II cepheid variable having the same period of light variation. The space distributions of the type II cepheids and RR Lyrae stars are also similar, and it is assumed that these objects are closely related.

MIRA-TYPE OR LONG-PERIOD VARIABLES. Most numerous of the variable stars are the long-period or Mira-type variables. These objects are very cool red giants and supergiants with periods of light variation that range from 70 to more than 1000 days. However, their periods are not constant like those of the cepheids, and maximum or minimum light can occur weeks before or after the dates predicted from the mean light curve. Because the temperatures of the Mira variables are often so low (less than 2500°K), they tend to radiate most of their energy at the invisible infrared wavelengths at the low-temperature phases of their cycles. Thus the star appears to vary greatly in brightness in the visible range not only because of changes in overall luminosity but also because the region of highest energy output shifts back and forth between the visible and the infrared. Thus o Ceti, or Mira, the prototype of this class of star, can vary in visual brightness from a prominent 2nd- or 3rd-magnitude star to a 9th-magnitude star that can be viewed only with optical aid. The Mira variables also display a complicated series of changes in spectra during their light cycles which are not yet fully understood.

SEMIREGULAR VARIABLES. Closely related to the Mira variables, the red semiregular variables exhibit a similar overall periodicity but display short-term unpredictable changes. The range of light variation in the semiregulars also tends to be smaller than that of the Miras.

IRREGULAR VARIABLES. Irregular variable stars exhibit no periodicity whatsoever. Their observed properties, such as their color and spectra, strongly suggest that they are related to the Miras and the semiregulars.

T TAURI STARS. The T Tauri stars, usually found in star clusters and associations having high gas and dust content, also exhibit short-term irregular variations in brightness. Their spectra are an unusual combination of emission lines of hydrogen and metals on an odd background continuum that suggests an object intermediate between stars and nebulosity. It is believed that the T Tauri stars represent a protostar stage of a star's development and that the light variations reflect anomalies in the gravitational contraction process.

The H-R diagram for intrinsic variable stars is shown in Fig. 10.3.

Eruptive Variables. In addition to the pulsating variables, there are stars called eruptive variables that exhibit sudden and erratic outbursts of light. Often such outbursts of visible light are accompanied by ejections of gas shells and bursts of radio radiation.

FLARE STARS. At the lower end of the main sequence, located among the M dwarfs, are the flare stars or UV Ceti stars. These objects can increase their brightness by several magnitudes over a matter of minutes. During these outbursts astronomers have observed radio emission as well as sudden changes in spectra that are reminiscent of the behavior of solar flares; indeed, the total energy associated with these events, about 10^{30} ergs/sec, is comparable to that emitted by the largest solar flares. The available evidence thus indicates that these events are probably flare-type phenomena that, because of the low

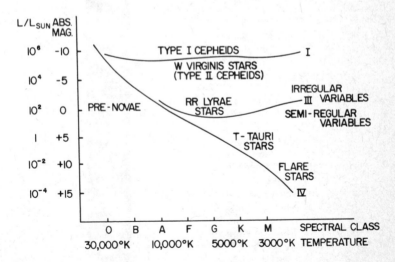

Fig. 10.3 The H-R diagram for intrinsic variable stars.

luminosities of the M-dwarfs, can be observed at interstellar distances.

NOVAE AND SUPERNOVAE. Most spectacular of all the variable stars
are the gigantic stellar explosions called novae and supernovae.
Beginning as hot, small-sized stars below the upper main sequence,
novae over a matter of a few hours or days increase their luminosity
by 6 to 10 magnitudes to attain an absolute magnitude as high as -10,
then over weeks or months fade to their original brightnesses. Novae
also eject gaseous shells into space at velocities of more than 1600
km/sec; these shells can be detected spectrosopically (see Fig. 10.4).
Months or years after the outburst, an expanding shell of nebulosity
can often be detected photographically (see Fig. 10.5).

Supernovae are much rarer than novae and can attain absolute magnitudes of -20, comparable to the total energy output of an entire
galaxy! The most famous supernova occurred in 1054 and was recorded by Chinese observers. The vestiges of this event are believed to
be visible today as the Crab nebula in Taurus, a complicated physical
system that is a strong source of radio waves and contains a small, extremely compact stellar remnant called a pulsar (see p. 164).

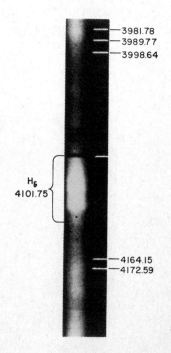

Fig. 10.4 The spectrum of a nova shell (Nova Aquilae, 1918) shortly after initial outburst. Along the line of sight to the nova we are viewing the explosion through a layer of gas that is rushing toward us at very high negative radial velocity, hence the absorption feature at the violet edge of the line profile for Hδ. Away from the line of sight the gas shell is not moving toward us as rapidly and we also see these parts of the shell as emission features, hence the broad emission longward of the violet edge absorption feature for the Hδ line. (Courtesy of the Hale Observatories.)

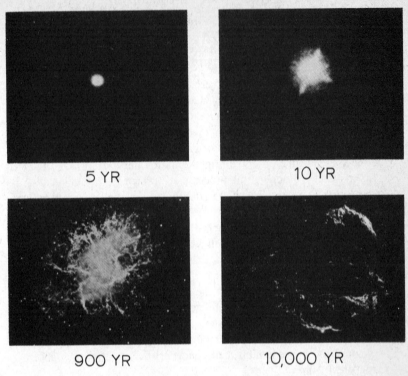

5 YR

10 YR

900 YR

10,000 YR

Fig. 10.5 Progression of a nova shell. (Courtesy of the Hale Observatories.)

P CYGNI STARS. The P Cygni stars are among the hottest of the eruptive variables and typically possess O and hot B-star spectral characteristics. Their light variations are slow and erratic, but qualitatively resemble those of novae. The P Cygni stars also exhibit weak emission lines in their spectra, and it is believed that they are ejecting material into the interstellar medium in the form of gas shells.

STARS OF UNUSUAL STRUCTURE

Although most stars have an interior and atmospheric structure much like that of our sun, there are stars whose structures differ markedly from those of the main sequence. Through careful studies of these structurally different types, astronomers have gained insight into the dynamics involved in a star's evolution.

Stars with Extended Atmospheres. Stars with expanding gas shells are said to have extended atmospheres. These are the shell stars, B-emission stars, Wolf-Rayet stars, and the so-called planetary nebulae.

SHELL STARS. Among the B stars are objects whose spectra display bright-emission lines of hydrogen and narrow dark-absorption lines that indicate extremely rapid rotation. It is thought that because of this rapid rotation, material has been ejected from the equatorial region of the star, forming rings or shells of gas about it. Such gas shells would be the source of the observed bright-emission lines. The study of such stars may help to unravel the processes by which mass loss occurs.

BE OR B-EMISSION STARS. In addition to the shell stars, there are about 4000 class B stars whose spectra exhibit emission lines of hydrogen and possibly other elements as well. These stars are called Be or B-emission stars, and it is believed that, as in the case of the shell stars, the characteristic emission lines arise from ejected material. Be stars are generally found to have extremely high rotation rates, and astronomers have speculated that the mass ejection and rapid rotation rates in these stars are directly related, as they are in shell stars. The chief difference between these stars and the shell stars is that the latter contain a significantly greater amount of material in their surrounding envelopes and hence exhibit their characteristic sets of narrow-absorption "shell" lines.

WOLF-RAYET STARS. Occupying a lofty position in the hierarchy of stellar surface temperatures are the Wolf-Rayet stars, formerly considered a subclass of the O stars but now classified as W stars. The Wolf-Rayet stars exhibit broad-emission lines, usually of nitrogen or carbon. On the violet edge of each emission line is a sharp absorption line not unlike that observed for novae and indicative of the presence of a rapidly expanding gaseous shell. Surface temperatures of Wolf-Rayet stars have been estimated to be as high as 100,000°K.

PLANETARY NEBULAE. The so-called planetary nebulae are the largest of the stars with extended atmospheres. They consist of a hot, central star surrounded by a slowly expanding gaseous shell that can be as large as a light-year across (see Fig. 10.6). Some are observable telescopically as small planet-like disks of light, from whence comes their name. The temperatures of the central stars of the planetary nebulae rival or even exceed those of the Wolf-Rayet stars, occasionally reaching values of well over 100,000°K. However, the planetary nebulae display much less violence in their dynamics than do the novae and

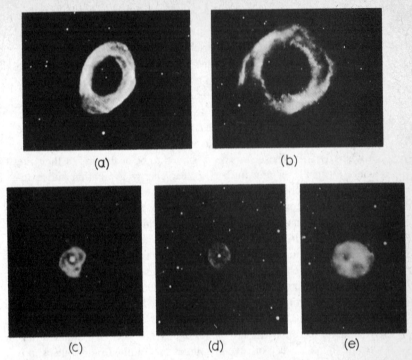

Fig. 10.6 Five planetary nebulae: (a) the Ring nebula M57 in Lyra, (b) NGC 7293 in Aquarius, (c) NGC 6543 in Draco, (d) I 3568 in Camelopardus, and (e) the Owl nebula M97 in Ursa Major. [(a), (b), (d), (e) Courtesy of the Hale Observatories; (c) Lick Observatory photographs.]

Wolf-Rayet stars, although they appear to have ejected much greater amounts of material into their surrounding shells.

RED GIANTS. The normal red giants are stars of spectral classes ranging from late G to late M and of high luminosities that place them well above and to the right of the main sequence of the H-R diagram. Although red giants do not exhibit any significant element abundances or light variations, they do possess highly tenuous and extended atmospheres that reach hundreds of millions of kilometers into space. For example, if the red supergiant Antares in the constellation Scorpius were placed at the center of the solar system, all the planets out to and including Mars would be orbiting within the confines of its atmospheric layers.

Compact Stars. In contrast to stars with extended atmospheres are the compact stars, whose masses occupy almost inconceivably small volumes of interstellar space.

WHITE DWARFS. In 1844 Friedrich Bessel of Germany discovered a wobbling effect in the motion of the star Sirius and correctly attributed it to the gravitational effect of an unseen companion of about one solar mass. In 1862 the predicted companion was observed visually and found to be a hot, but surpringly faint star having a radius about that of the earth. Many objects similar to Sirius's companion have since been discovered and constitute an important class of stars known as the white dwarfs. These stars are assumed to be at the end of their life cycles, incapable of any further nuclear energy generation and compacted by an unchecked gravitational collapse.

PULSARS. Theorists have long speculated on the possibility of a stellar gravitational collapse of such magnitude that it could crush the protons and electrons of a star's atoms into neutrons. Such a collapse would produce an object even more compact and dense then a white dwarf. Under such conditions the electrons and protons in the star are gravitationally crushed and compacted into neutrons. Most astronomers thought that if such objects, called *neutron stars*, existed, they would be difficult to detect at interstellar distances. In 1967, however, radio astronomers at Cambridge University in England announced the discovery of several sources that emitted radio pulses in a

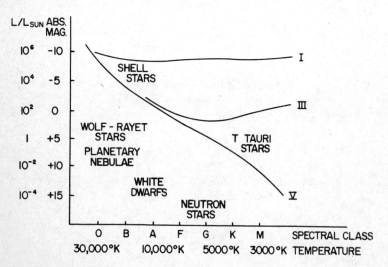

Fig. 10.7 The H-R diagram for stars of unusual structure.

highly regular fashion, with periods from $^1/_{30}$ to more than 3 seconds. These objects were named pulsars. In light of recent theoretical and observational investigations, it appears that pulsars are indeed compact neutron stars that are spinning on their axes every few seconds. The explanation for the intense, periodic radio beams emanating from them is at present a subject for debate among astrophysicists.

The H-R diagram for stars of unusual structure is shown in Fig. 10.7.

REVIEW QUESTIONS

1. Describe the spectra of the S stars, carbon stars, and barium stars. To what typical stars do these objects' temperature and luminosity correspond?
2. How can a cepheid variable be used to determine distances?
3. Compare and contrast Mira variables, semiregular variables, and irregular variables.
4. What are the differences between a nova and a flare star?
5. Calculate the density of a white dwarf having the mass of the sun and a radius equal to that of the earth. *Ans.:* mean density $= 1.5 \times 10^5$ g/cm^3.
6. What are pulsars?

11

The Interstellar Medium

The variety of material in the interplanetary medium strongly suggests that the space between the stars, the interstellar medium, may be similarly occupied. Although this view was strongly supported in the last century by the discovery of great nebular clouds along the Milky Way plane, it was not until well into the present century that astronomers had the necessary instrumentation to investigate the interstellar medium. Radio telescopes, photoelectric equipment, and a number of other devices have revealed the presence of both gas and dust in the vast space between the stars.

INTERSTELLAR GAS

Gaseous matter is the principal component of the interstellar medium. However, its density is low, about 10 atoms/cm^3, constituting a better vacuum than can be produced in a laboratory on earth.

Evidence for Interstellar Gas. The presence of gas in the interstellar medium is indicated by emission lines from nebulae, superposed absorption lines on the spectra of hot stars, and radio emission lines.

EMISSION LINES FROM NEBULAE. The most convincing evidence for the presence of interstellar gas are the dozens of bright patches of

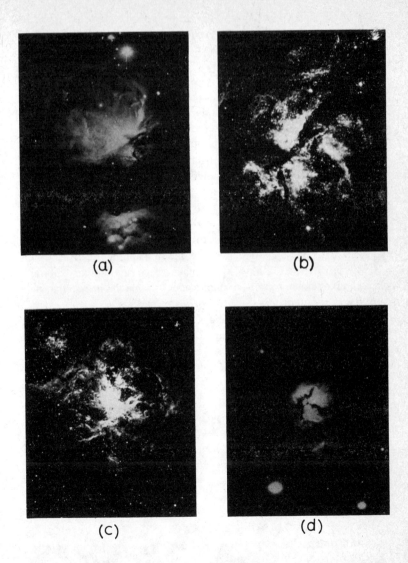

Fig. 11.1 Some gaseous nebulae: (a) the Great Nebula in Orion, (b) the Eta Carina nebula, (c) the Tarantula nebula in the Large Magellanic Cloud, and (d) the Trifid nebula in Sagittarius. [(a) Lick Observatory photographs; (b) the Cerro Tololo Inter-American Observatory; (c) by permission of the Harvard College Observatory, Cambridge, Mass.; (d) courtesy of the Hale Observatories.]

nebulosity that line the plane of the Milky Way and whose delicate structure can be seen only on long-time exposure photographs (see Fig. 11.1). The gaseous nature of these nebulae is confirmed by the presence of bright-line spectra.

SUPERIMPOSED ABSORPTION LINES. The detection of absorption lines (Fig. 11.2) superimposed on the spectra of hot stars is another indication of the presence of interstellar gas. Such lines would not arise in the atmosphere of a hot star. Moreover, the radial velocities of these lines are significantly different from those of the star, thus precluding the possibility that the lines arise from the second component of a binary star or a shell of gas surrounding the star.

RADIO EMISSION LINES. In recent years radio astronomers have discovered an impressive array of emission lines that arise from low-energy downward transitions of various atoms and molecules. The most important of these is the 21-cm line produced by neutral atomic hydrogen. This line is particularly valuable in mapping the distribution of material in the Milky Way plane because, unlike shorter-wavelength photons, the 21-cm photon can traverse the interstellar medium relatively undisturbed for thousands of parsecs.

Composition of the Interstellar Gas. The most abundant element in the interstellar gas is hydrogen. Helium is also present, and trace amounts of heavier substances. Hydrogen atoms are found in both neutral and ionized states. Regions of neutral hydrogen atoms are

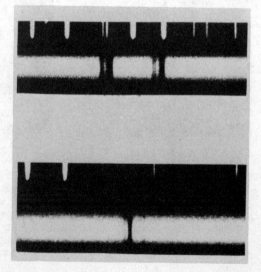

Fig. 11.2 Interstellar absorption lines of sodium (top) and calcium (bottom). Note the component lines due to several clouds of material. (Courtesy of the Hale Observatories.)

referred to as *H I regions* and regions of ionized hydrogen, *H II regions*.

INTERSTELLAR ATOMS. Because of the cold temperatures that pervade the interstellar medium (10–50°K), only a few elements are in an appropriate state of excitation to be spectroscopically detectable. Thus, studies conducted on interstellar absorption lines indicate the presence only of sodium, potassium, calcium, titanium, and iron in very low densities (less than 1 atom/cm^3). Because the physical conditions along the line of sight are not well understood, accurate abundance determinations involving such lines are difficult to make. Abundance estimates for these elements made on the basis of the strength of the absorption lines are similar to the relative abundances of these elements found from spectral analyses of stellar atmospheres.

Interstellar gas near hot stars is excited and ionized by short-wavelength continuum radiation from the star. As the ions recombine, photons are emitted and the gaseous region fluoresces or glows dimly, much as comets in the vicinity of the sun give off a dull glow. The bright, diffuse nebulae thus exhibit a weak continuous spectrum on which are superimposed a number of emission lines of varying strengths. Most of these lines are easily found to correspond with known elements such as hydrogen, helium, and neon. In the early part of this century, however, several of the observed emission lines eluded identification and were attributed to an unknown element called *nebulium*. Nebulium, however, suffered the same fate as coronium, the hypothetical element on the sun, when Ira Bowen at Mount Wilson discovered that at the low densities existent in the gaseous nebulae, atoms could deexcite themselves by downard or "forbidden" radiative transitions out of low-lying long-lived excitation states. Under normal conditions, these levels, called metastable levels, are deexcited by collisions with other atoms, but at low nebular densities, atomic collisions are so infrequent that the forbidden radiative emission has an opportunity to occur. With this insight, Bowen was able to show that the nebulium lines were the result of forbidden downward transitions in oxygen, nitrogen, and neon. A knowledge of the approximate physical conditions in a given nebulosity permits estimates to be made regarding the composition of the nebula based on the strengths of the emission lines present in its spectrum. Such studies indicate that the abundances in nebulae are very similar to those of the stars embedded in them, a fact that strongly suggests a close connection between the two.

One of the more interesting aspects of the interstellar gas is the presence of the high-energy atomic particle flux known as cosmic radiation or cosmic rays. As noted in Chapter 7, these particles probably arise from eruptive events such as stellar flares, novae, or supernovae explosions and then are accelerated by the galactic magnetic field. The element abundances in the interstellar cosmic ray flux, unlike those in the other observable aspects of the interstellar gas, are found to be significantly different from the normal abundances. For example, lithium, beryllium, and boron are more abundant in cosmic rays than in stellar atmospheres. The cosmic-ray abundance anomalies are considered to be important clues to the origin of these mysterious particles, but the cosmic-ray puzzle still defies solution.

INTERSTELLAR MOLECULES. More surprising than the existence of atoms of simple elements in the interstellar medium is the wide variety of low-abundance interstellar molecules that are found (see Table 11.1). Radio astronomers in the last decade alone have discovered emission lines emanating from more than twenty molecular substances

Table 11.1. Some Interstellar Atoms and Molecules

Interstellar Atoms		Interstellar Molecules	
Element	Symbol	Molecule Name	Formula
Hydrogen	H	Acetaldehyde	CH_3CHO
Helium	He	Ammonia	NH_3
Lithium	Li	Carbon monosulfide	CS
Beryllium	Be	Carbon monoxide	CO
Boron	B	Carbonyl sulfide	OCS
Carbon	C	Cyanogen radical	CN
Nitrogen	N	Ethyl alcohol	C_2H_5OH
Oxygen	O	Formaldehyde	H_2CO
Fluorine	F	Formic acid	$HCOOH$
Neon	Ne	Hydrogen molecule	H_2
Sodium	Na	Hydrogen cyanide	HCN
Magnesium	Mg	Hydrogen sulfide	H_2S
Aluminum	Al	Hydroxyl radical	OH
Silicon	Si	Isocyanic acid	$HNCO$
Sulfur	S	Methyl alcohol	CH_3OH
Chlorine	Cl	Methyl cyanide	CH_3CN
Argon	Ar	Methylacetylene	CH_3CCH
Potassium	K	Methylidyne	CH
Calcium	Ca	Silicon monoxide	SiO
Titanium	Ti	Thioformaldehyde	H_2CS
Iron	Fe	Water	H_2O

in the millimeter and centimeter wavelength range, and there are undoubtedly many more as yet undiscovered. The compounds range in complexity from molecular hydrogen (H_2) to simple hydrocarbons such as ethyl alcohol (C_2H_5OH). Observations of these molecules indicate that they are strongly associated with known clouds of gas and dust in the Milky Way, and their origin may be linked to the physical conditions existent in these objects.

Theories for the origin of the interstellar molecules include formation by collisions of atoms in the interstellar medium; formation in the atmospheres of cool stars or protostars such as the carbon stars, Mira variables, or globules; and formation on the surfaces of dust particles that would act as "atom collectors."

Distribution of Gas. Even a casual glance at photographs of the Milky Way reveals that the distribution of interstellar gas is not uniform. Supporting evidence to this effect is provided by the fact that the interstellar lines, both absorption and emission, often display several components that can be accounted for only by assuming that each component line arises from a gas cloud at a different distance from the earth and moves in a different orbit about the galactic center than that of the sun, thus producing a unique radial velocity relative to the sun. Additional information concerning distances to nebular clouds can be obtained from spectroscopic parallaxes of stars contained within the nebulosity. Such studies of the distribution of interstellar gas clouds indicate that the interstellar gas tends to collect in large, curving spiral arms dozens of parsecs across (see Chapter 14).

INTERSTELLAR DUST

Approximately 1 percent of the mass of the interstellar material consists of tiny grains of dust, about $1/10$ micron in diameter. This works out to one dust particle for every 10^{12} gas atoms in the interstellar medium.

Evidence for Interstellar Dust. Evidence for the presence of dust in interstellar space is provided by dark nebulae, reflection nebulae, reddening of starlight, and polarization of starlight.

DARK NEBULAE. Along the Milky Way plane can be seen a number of dark areas, the largest being the Great Rift in the Ophiuchus-Cygnus regions and the famous Coalsack nebula. These dark nebulae, as they are called, are enormous clouds of material that can dim or blot out the light from any stars or nebulosity that lie behind them (see

Fig. 11.3). The particle size required to absorb and thus obscure the starlight must be significantly larger, about 10^4 to 10^5 times, than the atoms and molecules of the interstellar gas, given the known densities of the latter. Thus astronomers have deduced the existence of a second component of the interstellar medium, the interstellar dust.

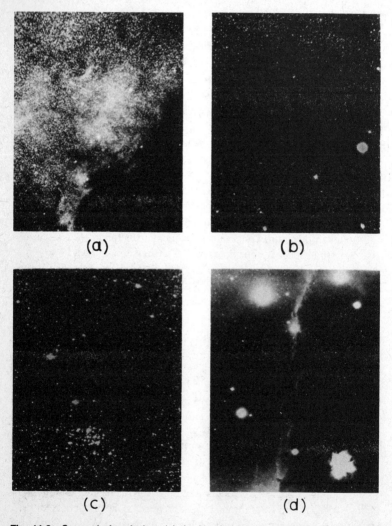

Fig. 11.3 Some dark nebulae: (a) the North America nebula in Cygnus, (b) an S-shaped nebula in Ophiuchus, (c) Barnard 86 Sagittarii, and (d) the Horsehead nebula in Orion. (Courtesy of the Hale Observatories.)

Fig. 11.4 The reflection nebulosity about the Pleiades star Merope. (Courtesy of the Hale Observatories.)

REFLECTION NEBULAE. Some bright, diffuse nebulae have spectra that do not show the emission lines of the gaseous nebulae, but rather display the same spectral characteristics as the stars embedded in them (see Fig. 11.4). This observation can be explained only by assuming that the nebulae are reflecting, or scattering, the light from the embedded stars. Since interstellar gas clouds do not have the densities needed to produce such reflection, astronomers have assumed that the observed reflection is caused by larger-sized particles.

REDDENING OF STARLIGHT. Extinction of starlight by both absorption and scattering is thus evidence of the existence of dust-sized particles in the interstellar medium. Another indication is the reddening of starlight due to selective scattering of blue, or short-wavelength, light. If a hot O or B star at a considerable distance from earth is observed spectroscopically and photoelectrically, it appears much redder than would be expected for its spectral type. This difference between the intrinsic color and the observed color is called the color excess. The photographic absorption (r) or dimming of a star's light from the effect of dust particles is, on the average, equal to $3.0 \times CE$, where both the color excess CE and the absorption (r) are expressed in magnitudes. The existence of color excesses and absorption can be accounted for only by the presence of particles larger than atoms or molecules.

POLARIZATION OF STARLIGHT. Dust in the interstellar medium can polarize light from the stars. As stars of greater and greater distance are observed, the light becomes increasingly polarized; that is, it has more and more of a single orientation for its vibrational plane. Moreover, the plane of polarization is roughly perpendicular to the plane of the Milky Way. Such a result cannot be accounted for in terms of atom-sized particles in the interstellar medium.

Nature of the Dust Grains. Unlike the gases in the interstellar medium, the dust grains do not lend themselves to spectroscopic analysis, so information about them must be obtained indirectly. The phenomena of reddening and polarization of starlight both indicate something of their nature.

It is almost certain that the dust grains are significantly larger than atoms or molecules. If the grains possessed atomic dimensions, their ability to scatter short wavelengths of light should be considerably higher than what is observed. On the other hand, micron-sized particles scatter short-wavelength radiation to much less a degree than do atoms or molecules and hence more closely fit the observed behavior of the interstellar grains. The manner in which starlight is polarized as it passes through dust clouds suggests that the dust grains are elongated, like interstellar needles, and roughly aligned, probably with interstellar magnetic fields, perpendicular to the plane of the Milky Way. The composition of these grains is still uncertain, but they are probably a mixture of carbon and frozen compounds of hydrogen.

Distribution of Dust. Distances to the gaseous and reflection nebulae are obtained from the spectroscopic parallaxes of the embedded stars; distances to the dark nebulae are determined by the use of the

Wolf diagram. The Wolf diagram is basically a plotting of the number of stars visible in a region of sky down to an apparent magnitude *m* versus the value of *m*. The star counts are conducted on two regions of the sky, one of which is within the dark nebula of interest, and the other in a nearby, unobscured star field as shown in Fig. 11.5. Out to the nebula the two star count curves are virtually coincident. At the near edge of the dust cloud, the stars in the cloud become progressively more obscured by the cloud's ever-thickening layers as the counts proceed to fainter magnitudes. Thus, the star count curve for the cloud region diverges from that for the outside region. The divergence continues until the far edge of the cloud. At this point, the

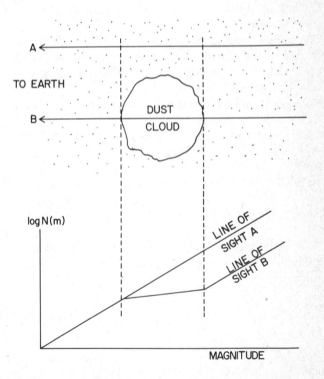

Fig. 11.5 An idealized Wolf diagram. Because of the cloud's absorption, the star counts measured along path B will be different from those measured along the unobscured path A. By measuring the exact nature of this departure, astronomers can deduce not only the cloud's distance but other properties of the cloud as well.

star counts for the cloud region once more increase at the same rate as for the outside region. By noting the magnitude m_t at which the curves diverge and assuming a mean absolute magnitude $\overline{M}$ for all of the region's stars, an estimate of the distance modulus $m_t - \overline{M}$ and thus the distance to the dust cloud's lead edge can be obtained.

Such studies indicate that although the general distribution of the dark dust clouds of the Milky Way is similar to that of the gaseous nebulae, significant and puzzling differences exist.

REVIEW QUESTIONS

1. Summarize the evidence for the existence of interstellar gas.
2. List at least five elements that exist in the interstellar medium. How are they detected?
3. Compare and contrast the processes that give rise to the interstellar spectral lines formed in deep space away from stars and those that give rise to the spectra of gaseous nebulae. Include in your discussion the spectra of reflection nebulae.
4. List several molecules known to exist in the interstellar medium. How are they detected?
5. Summarize the evidence for the existence of interstellar dust.
6. What is known about the nature of the dust grains?
7. An A0 star is observed to have a color excess of 1 magnitude. If the observed apparent magnitude of the A star is + 13, find the distance to the star corrected for interstellar absorption if both the absolute magnitude and intrinsic B-V color for the AO star are zero. *Ans.*: 10^3 pc.
8. Describe the workings of a Wolf diagram. How can it be used to determine the distance to a dust cloud?

12

Stellar Evolution

Stars, like life forms, undergo changes in physical properties, but over a vast period of time. Thus, astronomers are faced with the problem of deducing the life cycles of the stars from the seemingly static conditions they observe. However, by constructing stellar models from observational data with the help of high-speed computers, they have been able to outline roughly the various phases of stellar evolution and to account for a wide variety of structures and forms in an all-encompassing theory.

STAGES OF STELLAR EVOLUTION

The evolutionary path a star takes is determined primarily by its mass and the gravitational forces that act upon it. In fact, once a star is formed, its life cycle can be thought of as a duel between gravity, which attempts to draw the gas into an ever more compact sphere, and the release of energy, which seeks to blow the sphere apart. The more massive the star, the shorter its lifespan.

Beginning Stages. It is almost certain that stars form from the gas clouds of the interstellar medium (see Chapter 11), but what initiates this process is not known. One possibility is that interstellar gas clouds of different temperatures collide and the gas atoms condense much as water vapor forms on the earth when warm and cold air masses meet. Once condensation has begun, the *protostar* contracts under gravity and shines dimly by the conversion of its gravitational potential energy into electromagnetic energy. At this stage, the protostar can often be seen as a globule or dark spot in photographs of gas clouds such as the Rosette nebula (Fig. 12.1). As contraction proceeds, the star moves almost straight downward on the H-R diagram along an evolutionary path called a *Hayashi line* or track. Finally, the internal temperature and pressure of the protostar increase to the point where thermonuclear reactions can take place. The additional energy from the nuclear reactions is sufficient to halt the gravitational collapse of the protostar, and the resulting balance of forces (*hydrostatic equilibrium*) maintains the protostar's radius at a constant value. The protostar is now said to have reached the main sequence and is, in fact, a full-fledged star.

Two classes of objects, the T Tauri stars and the Herbig-Haro ob-

Fig. 12.1   Interstellar globules in the Southern Milky Way. (By permission of the Harvard College Observatory, Cambridge, Mass.)

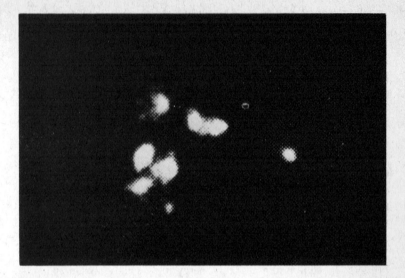

Fig. 12.2 Herbig-Haro objects. (Lick Observatory photographs.)

jects, have been observed that may be protostars in evolutionary stages intermediate between globules and the main sequence. The Herbig-Haro objects (Fig. 12.2) are bright spots sometimes observed in dark nebulae and are thought to be protostars at or just before the stage at which nuclear reactions in the core begin. T Tauri stars (see Chapter 10) seem to represent the last stage of a protostar prior to its becoming a main sequence star. The time for a star to contract from an interstellar gas cloud to a main sequence star has been calculated to be 10^5 years for a 20 solar mass star, 10^6 years for a 1 solar mass star, and 10^8 years for a 0.1 solar mass star.

The Main Sequence Stage. The greater part of a star's lifetime is spent in the relatively stable, main sequence stage of its evolution; indeed, most of the stars on the H-R diagram are found there. The star's final position on the main sequence is determined almost solely by its mass. The most massive stars are the hottest and most luminous and occupy the upper main sequence; the least massive stars are low in temperature and luminosity and lie on the lower main sequence. The less massive protostars are much more likely to form than the more massive ones, but massive protostars evolve at a much faster rate. Thus a protostar about 100 solar masses may contract into a main sequence star within 100,000 years, whereas a 0.05 solar mass protostar might take over a billion years to reach the main sequence stage. Typical lifetimes for stars on the main sequence range from several million years for stars of 20 solar masses to over 10 billion years for those of 0.1 solar mass. Our sun is estimated to have a lifetime of about 10 billion years on the main sequence, of which about 5 billion years has been exhausted.

The Red Giant Stage. Although the outward physical characteristics of a star do not change appreciably while it is on the main sequence, changes are occurring in its core. Hydrogen fuses into helium, and the heavier helium atoms sink to the center of the star to form a helium core surrounded by a hydrogen envelope. The helium core grows in size and eventually reaches the point at which it begins to collapse on itself gravitationally. As it does so, it releases gravitational potential energy, which heats the hydrogen at the core boundary. The fusion rate of hydrogen, which is highly sensitive to temperature, increases markedly, as does the rate at which the star produces energy. The equilibrium between energy release and gravitational collapse is thus upset; the outer layers of the star then begin to expand and the total luminosity of the star increases. Since the star's surface temperature is directly proportional to the square root of the luminosity and inversely proportional to its radius, as the radius increases in size, the effect of the increased luminosity is offset, and the surface temperature decreases. The star therefore leaves the main sequence and moves up and to the right to the region of increased luminosity and decreased surface temperature; this region is called the red giant branch. Ultimately the temperature and pressure in the helium core reach the point called the *helium flash* at which the helium fusion reaction, or *triple-alpha process,* is initiated. The ignition of helium fusion, in which three helium atoms are fused into a carbon atom, marks the end of the red giant phase, a stage that has a duration of some 10^7 years for a 20 solar mass star, 2×10^9 years for a 1.0 solar mass star, and more than 10^{10} years for a 0.1 solar mass star.

Post Red Giant Phase. Astronomers can only hypothesize about the evolutionary processes that occur in the star after the red giant phase. However, it is generally accepted that the star goes through rapid and unstable phases during which its nuclear energy is exhausted. If the star is significantly larger than 1 solar mass, it is believed to shed much of its mass by means of nova or supernova explosions or by the less catastrophic method of throwing off shells or rings of gas. Thus by the end of the post red giant phase the star is

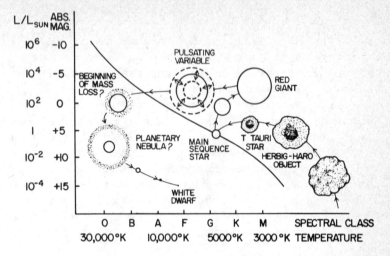

Fig. 12.3 The evolutionary path of a star of about 1 solar mass.

pared down to 1 solar mass or less and is no longer capable of generating nuclear energy (see Fig. 12.3). The expelled mass returns to the interstellar medium where it may later contract into a star.

Final Evolutionary Stages. Stars in the final stages of their evolutionary cycles constitute some of the most interesting objects in the universe. With the nuclear processes exhausted, there is nothing to prevent a star from contracting into an exceedingly dense configuration. Depending on the star's mass as it enters this final evolutionary stage, three possible "superdense" objects are possible: the white dwarf, if the remaining mass is less than 1.4 solar masses; the neutron star, if the remaining mass is between 1.4 and 3 solar masses; and possibly a black hole, if the remaining mass is larger than 3 solar masses.

WHITE DWARFS. White dwarfs are stars that have collapsed to the extent that 1 solar mass is packed into a sphere the size of the earth; 1 cubic centimeter of white dwarf material, which is mostly carbon, may weigh several thousand kilograms. The surface temperatures of these objects are still relatively high, but they have low luminosities because of their small size and the absence of nuclear energy. Thus white dwarfs glow like embers of a dying fire. Eventually they cool into the cold, dark masses referred to as *black dwarfts*. White dwarfs, however, are this stable only if they are of less than about 1.4 solar masses. Theoretically, the gravitational collapse of a white dwarf will continue unchecked if its mass exceeds this critical value, called the *Chandrasekhar limit*.

NEUTRON STARS, OR PULSARS. The more massive stars eject much of their mass in nova or supernova explosions. The core that remains after such an explosion is a highly compact sphere that astronomers observe as a neutron star, or pulsar (see Chapter 10).

BLACK HOLES. Astrophysicists have predicted a more extreme state of gravitational collapse in which 1 solar mass is compressed into a sphere less than 3 km in diameter. For such objects, called black holes, the escape velocity would be greater than the velocity of light. Thus, photons could not escape, and the black holes could not be detected electromagnetically. Some investigators have attributed fluctuations in the strength of X rays from certain sources to the movements of a star's massive but invisible companion—a black hole. However, evidence for the existence of black holes at this writing is not conclusive.

EVIDENCE FOR STELLAR EVOLUTION

The study of star clusters and stars of different population types provides evidence for the current theory of stellar evolution.

Star Clusters. Perhaps the best evidence that stars have life cycles similiar to that outlined above comes from the star clusters (see Chapter 13). Clusters may contain as few as five or six stars or as many as 1 million. All the stars of a cluster are formed out of one large cloud of matter at approximately the same time. Because the dimensions of a cluster are negligible compared to the distance to a cluster, all of the cluster members can be regarded as being at the same distance from earth. As a result, a plot of apparent magnitude versus the color for the stars in the cluster (called a color-magnitude diagram) is in essence an H-R diagram because the correction of the apparent magnitude for distance will be the same for all of the cluster stars. For a recently formed star cluster, only the more massive stars have had a chance to reach the main sequence. A color-magnitude diagram constructed for such a cluster will be similar to that shown in Fig. 12.4a. A more evolved cluster will have a more developed main sequence into the giant branch (Fig. 12.4b). In the oldest and most evolved clusters (Fig. 12.4c), only the very low-mass, slow-evolving stars are found on the main sequence; the giant branch is well populated with all of the cluster stars that have left it.

The idealized composite main sequence formed from the main

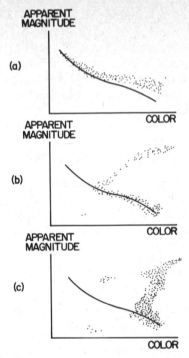

Fig. 12.4 The color-magnitude diagram for a hypothetical (a) young star cluster, (b) intermediate-age cluster, and (c) old cluster.

sequence location of the stars of young, middle-aged, and old clusters is referred to as the *zero-age main sequence*. The zero-age main sequence can thus be thought of as a sequence of positions on the H-R diagram to which protostars will migrate just prior to the onset of nuclear processes in their cores.

Population Types. As the stars evolve through their life cycles, the material that they return to the interstellar medium is enriched with both helium and "metals" (the latter term is applied by astronomers to any substance heavier than helium). If this enriched material contracts into another star, that star will contain an even higher percentage of metals at the end of its cycle, and so on. Thus, one would expect to find the youngest stars highly metal-enriched, and the oldest stars metal-deficient. If one spectroscopically observes the stars believed to be in the oldest part of the Milky Way, in particular the halo and nucleus (see Chapter 15), one finds that these objects, called *Population II stars,* display the predicted metal-poor spectra, whereas those stars that are found in the young, spiral arm regions of the Milky Way, called *Population I stars,* exhibit relatively metal-rich spectra. The degree to which metals appear in a star's spectrum depends on the previous history of the material from which the star was formed. This history can vary considerably; thus stars have metal contents throughout the range between the metal-rich Population I stars and the metal-poor Population II stars.

ELEMENT FORMATION

The interiors of evolving stars are the greatest nuclear accelerators known, and it is here presumably that the building of elements through various types of nuclear reaction occurs. Eight basic nuclear reactions are advanced to account for the existence of the elements that are observed in the universe as well as their relative abundancies:

1. The fusion of hydrogen into helium:

$$4H^1 \rightarrow He^4$$

2. The fusion of helium into carbon (the *triple-alpha process*):

$$3He^4 \rightarrow C^{12}$$

3. The α *process,* in which helium nuclei react with carbon to form yet heavier elements such as oxygen, neon, magnesium, silicon, sulfur, argon, and calcium:

$$C^{12} + He^4 \rightarrow O^{16}$$
$$O^{16} + He^4 \rightarrow Ne^{20}$$
$$Ne^{20} + He^4 \rightarrow Mg^{24}, \text{ and so on.}$$

4. The *s process,* in which neutrons are produced and captured by a given nucleus at a sufficiently slow rate to allow the nucleus to β decay between successive neutron captures. For example, krypton is believed to be formed by the s-process reaction:

$$Br^{44} + n \rightarrow Br^{45} \rightarrow Kr^{45} + \beta$$

5. The *r process,* in which neutrons are produced and captured by a given nucleus at a sufficiently rapid rate that β decay does not occur between successive neutron captures. Antimony is thought to be formed by the following reaction:

$$Sn^{118} + n \rightarrow Sn^{119}$$
$$Sn^{119} + n \rightarrow Sn^{120} \rightarrow Sb^{120} + \beta$$

6. The *p process,* in which protons are captured by heavier nuclei such as in the reaction that produces xenon:

$$I^{123} + p \rightarrow Xe^{124}$$

7. The *equilibrium* or *e process,* in which the elements having atomic numbers close to that of iron are formed in the extremely hot cores of novae and supernovae. When the gas at the core suddenly changes its temperature and density as a result of the star's explosion, the elements present cannot react further and are hence "stuck" or "frozen" at the atomic numbers they possessed immediately prior to the star's outburst.

8. Some of the light elements including lithium, beryllium, and boron cannot be formed and maintained under the conditions that exist in the interiors of stars, yet these elements have been detected. To account for their presence, an unknown nuclear process, the *x process,* has been postulated, a process that would occur in the atmospheres of stars where the physical conditions are less extreme than those of the deep interior.

REVIEW QUESTIONS

1. Describe the various stages of stellar evolution. Why does a star proceed from one stage to the next?
2. Describe the role of gravity in stellar evolution.
3. Describe the role of nuclear processes in stellar evolution.
4. What evidence do astronomers have that stars do indeed evolve?
5. What are the nuclear processes by which the various elements are formed?
6. Propose a process by which fluorine, F^{19}, might be produced.

13

Star Clusters

Although the distribution of stars in space appears roughly uniform to the naked eye, collections of stars such as the Pleiades are readily discernible. The telescope reveals hundreds more of these aggregates, bound together by gravity. Because these clusters are composed of large numbers of stars whose lifetimes began at the same point in cosmic history, they provide considerable insight into the patterns and processes of stellar evolution.

Two types of star clusters have long been recognized: the *open* or *galactic clusters* found along the plane of the Milky Way, and the tightly knit *globular clusters* that are found on the Milky Way plane and in other parts of the galaxy as well. In the 1940s yet another type of cluster, the sprawling *stellar association,* was recognized (see Fig. 13.1).

GALACTIC CLUSTERS

The galactic clusters are by far the most numerous of the known star clusters. More than 800 are presently known, and there may be hundreds more in existence that have escaped detection because of their obscuration by the gas and dust clouds of the Milky Way. The stars in a galactic cluster are usually loosely packed, and the membership can range from less than 50 stars up to several thousand. Although their linear diameters rarely exceed 10 parsecs, galactic clusters are relatively close to the sun and often cover a considerable area of the sky. The likelihood is thus high that a given grouping will contain a certain number of field stars, that is, stars not in the cluster, but in the same line of sight as the cluster. The only recourse for the astronomer in such a case is to check both the proper motion and radial velocity of the suspected cluster member. If the motions of the object are the same as those of known cluster members, then it is highly probable that the object is indeed a member.

Distances. Distances to galactic clusters can be obtained by several methods, including the method of spectroscopic parallax, the use of the color-magnitude diagram, and the so-called moving cluster method.

THE METHOD OF SPECTROSCOPIC PARALLAX. In the spectroscopic parallax method, a star or stars of known luminosity are located in the

Fig. 13.1 Two star clusters: the great globular cluster M13 in Hercules (left) and the open or galactic star cluster M67 in Cancer. (Courtesy of the Hale Observatories.)

cluster and their apparent brightnesses measured. A comparison of the apparent brightness, which is usually corrected for interstellar absorption, and the known intrinsic brightness yields the distance to the cluster.

USE OF THE COLOR-MAGNITUDE DIAGRAM. In the color-magnitude method an observationally determined color-magnitude diagram main sequence is compared with the zero-age or theoretically unevolved main sequence from the H-R diagram. Because the H-R diagram plots absolute magnitude on the y axis, comparison with the cluster's color-magnitude diagram will yield the difference $m - M$, the value of the cluster's distance modulus, from whence the distance r, in parsecs, is obtained (see Fig. 13.2).

THE MOVING CLUSTER METHOD. If the star cluster is close enough to the earth, as is the Hyades cluster in the constellation Taurus, it may be possible to obtain a radiant point for the cluster motion very similar to that described for a meteor shower. The mean radial velocity of the cluster members can be obtained as well as the coordinates of

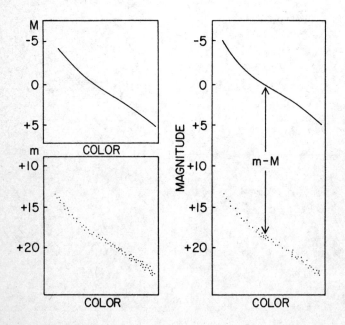

Fig. 13.2 The distance to a cluster. By comparing the cluster's main sequence with the zero-age main sequence (solid curve), the distance modulus $m - M$ and hence the cluster distance is obtained.

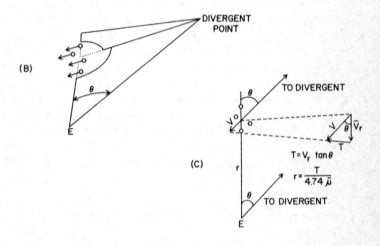

Fig. 13.3 The geometry of the moving cluster method: (a) side view, (b) oblique view, and (c) top view.

their geometric center and radiant point (see Fig. 13.3). From this information, astronomers can obtain the radial velocity component $\bar{V}_r$ of the cluster's motion as well as the angle between the mean radial velocity and the mean space velocity $\bar{V}$ of the cluster. Using these results, the mean tangential component $\bar{T}$ of the cluster's motion (in kilometers per second) can be calculated, from which the distance r (in parsecs) can be determined from the relation

$$r = \frac{\bar{T}}{4.74\bar{\mu}}$$

where $\bar{\mu}$ is the mean proper motion of the cluster members in seconds of arc per year.

Distribution. Once the distance to a galactic cluster having known celestial coordinates is obtained, its position in three-dimensional space can be readily determined. From such analyses, the galactic clusters are found to lie along the plane of the Milky Way (see Fig. 14.3), and as such, comprise an integral part of the galactic disk.

Age. If one assumes that all of the members of a star cluster were formed at the same time, then according to the present theories of stellar evolution (see Chapter 11), the most luminous, most massive main sequence stars will evolve off the main sequence first, to be followed by the progressively less massive, less luminous objects. Thus, the main sequence gradually disappears as the age of the cluster increases, as shown in Fig. 13.4. Using theoretical calculations regarding a star's lifetime on the main sequence (see Table 13.1), astronomers can estimate the age of a star cluster by simply noting the earliest spectral type at which the cluster's main sequence ends. By this procedure astronomers have deduced that the ages of galactic clusters range from a few million years for the extremely young clusters such as h and chi Persei to some 10 billion years for the galactic cluster M67 in Cancer. In support of this view is the observation that the clusters thought to be young systems are often associated with significant amounts of nebulosity, whereas the older clusters are devoid of

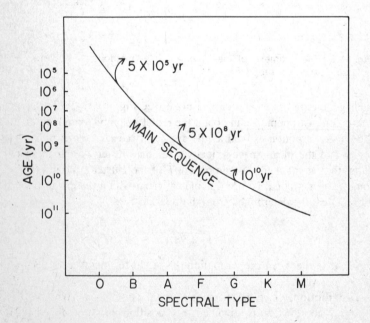

Fig. 13.4 Determining the age of a cluster. By noting the turn-off point of the main sequence, a cluster's age can be estimated.

Table 13.1

Spectral Type at Which Turn-off Point Occurs	Approximate Age of Cluster (years)
05	5×10^5
B0	5×10^6
B5	5×10^7
A0	5×10^8
F0	5×10^9
G0	1×10^{10}
K0	5×10^{10}
M0	1×10^{11}

interstellar material (see Fig. 13.1); presumably the material has either been used up in star formation or has been blown away by solar wind phenomenon. This method of age estimation can also be applied to stellar associations and globular clusters.

GLOBULAR CLUSTERS

Most impressive of all the star clusters are the 125 or so known globular clusters. These clusters resemble a bacterial culture when viewed or photographed through a telescope; thousands of individual stars are visible. Estimates of the actual numbers of stars present are difficult to make, owing to the close packing at the center (10^2 to 10^3 stars/parsec^3), but a globular cluster is thought to have several hundred thousand members, most of which are too dim to be detected from earth.

Distances. The globular clusters are so remote from earth that somewhat different methods of distance determination are used than those employed for the galactic clusters. Astronomers have discovered that W Virginis stars and RR Lyrae stars are often found among the stars comprising the globular clusters. Since the intrinsic absolute magnitudes of these stars can be determined from their pulsation periods, the distance to the globular cluster can be calculated from the distance modulus formula. Unfortunately, a significant number of globular clusters are so distant that individual stars are unresolvable and hence techniques using stellar distance indicators cannot be employed successfully. However, the angular diameters of such clusters can be measured, and by assuming an average linear diameter of about 50 parsecs for the cluster (a value based on data obtained for nearby globulars), the distance to more remote systems can at least be estimated.

Distribution. Distance determinations of the globular clusters indicate that these objects are at enormous distances, thousands of parsecs in some cases, from the sun. Early in this century the American astronomer Harlow Shapley established that globular clusters are flung throughout the outer regions or halo of the Milky Way in a roughly spherical distribution relative to the galactic center (see Fig. 14.3) and thus are not independent of the Milky Way system.

Age. The color-magnitude diagrams of globular clusters display virtually no main sequence and thus very closely resemble the hypothetical color-magnitude diagram shown in Fig. 12.4c. If the absence of the main sequence in the globular cluster is due to the stars' having evolved away, then the globular clusters are among the oldest objects in the universe, with ages of more than 10 billion years. This hypothesis is confirmed by the fact that globular clusters are metal-poor objects, which implies that the material out of which they were formed has not yet had the opportunity to undergo metal-producing nuclear reactions.

STELLAR ASSOCIATIONS

In the middle of this century astronomers recognized that there are certain classes of stars that are not relatively close together and yet are not randomly distributed across the sky. Investigations of the motions of these stars confirmed that these groupings, called stellar associations, are very loosely knit groups of stars ranging in size from 20 to 200 parsecs and containing between 10 and 100 stars. Membership in these systems is even more difficult to establish than in the case of the galactic clusters, owing to the large area of sky that is often covered by the association. For example, the Orion association, whose radiant lies in the constellation Orion, may have members as far away as the constellation Auriga.

Associations are generally composed of O and B stars, in which case they are called *OB associations,* or T Tauri stars, in which case they are called *T associations.* A few associations contain all these three types of stars and are thus combinations of OB and T associations. In such instances the association is denoted by the type of star that constitutes the majority of its membership.

Distances. Distances to associations are usually determined by measuring the spectroscopic parallaxes of their individual members or by the use of the moving cluster method described for the galactic clusters. Because of the great luminosity of the component O and B stars, associations at even greater distances than galactic clusters, often as far as 1000 or 2000 parsecs, can be detected.

Distribution. Like galactic clusters, associations lie along the Milky Way plane. In fact, associations are so closely allied with the gas and dust of the Galaxy that they have often been used to trace the galactic spiral structure.

Age. From the backward extrapolation of the motions of the associations' members, it has been deduced that these systems can be no more than a few million years old. This estimate is confirmed by the presence of large amounts of nebulosity and by the color-magnitude diagrams for these systems. Associations display either a well-developed upper main sequence as shown in Fig. 13.4a which contains O and B stars that have not yet had time to evolve off the main sequence to red giants and supergiants, or a series of T Tauri stars that lie above the lower main sequence and have not yet completed their contraction onto the main sequence. Because of their recent formation and their rapid evolutionary rates, associations are of particular interest to the study of stellar evolution.

REVIEW QUESTIONS

1. Compare and contrast the properties of globular clusters, galactic clusters, and associations.
2. An RR Lyrae star having a known absolute magnitude of 0.0 is observed in a globular cluster to have an apparent magnitude of +15. What is the distance to the cluster if the interstellar absorption is negligible? *Ans.:* 10^4 pc.
3. A galactic cluster is observed to have the following apparent magnitudes: A0 V, +5; G2 V, +10; M0 V, +15. How far away is this cluster? *Ans.:* 100 pc.
4. How are multiple stars similar to binaries? To star clusters? How are they different?
5. Describe how the age of a star cluster can be estimated.
6. Discuss three different methods by which the distance to a star cluster can be obtained. Which method(s) will work for the globular clusters? Explain.
7. Why are star clusters important to the study of stellar evolution?

14

The Milky Way Galaxy

With his primitive telescope, Galileo observed that the mysterious band of light girding the celestial sphere was actually a vast collection of individual stars. Not until this century, however, were astronomers able to make a reasonably accurate estimation of the magnitude and complexity of the Milky Way star system.

SIZE AND SHAPE OF THE MILKY WAY

If a composite photograph of the entire Milky Way is assembled as shown in Fig. 14.1, its planar symmetry can at once be recognized. The first attempts by Herschel·to investigate the size and shape of the Milky Way resulted in a "grindstone" model in which the sun was approximately at the center (see Fig. 14.2a). Early in the twentieth century, the Dutch astronomer Kapteyn, using more refined data, developed a model for the Milky Way that was ellipsoid-shaped with the sun at the center, as shown in Fig. 14.2b. Kapteyn also estimated the diameter of the Milky Way to be about 5000 light-years or just under 2000 parsecs. The present picture of the Milky Way (Fig. 14.2c), arrived at by Shapley and others, retains Kapteyn's ellipsoid-shaped nuclear bulge, but shows the sun to be some 10,000 parsecs or about two-thirds of the way from the galactic nucleus to the outer rim of a

Fig. 14.1 The Milky Way. (Lund Observatory, Lund, Sweden.)

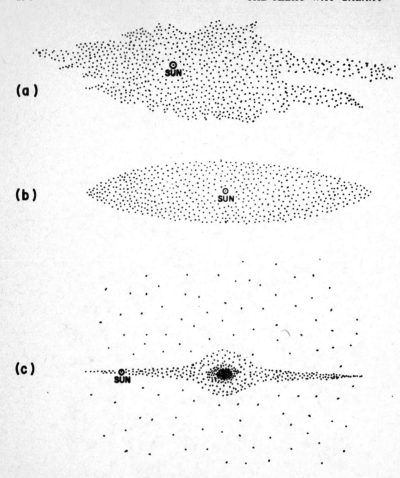

Fig. 14.2 Models of the Milky Way. (a) Herschel's model, (b) Kapteyn's model, and (c) present view.

disk of stars, gas, and dust that emanates from it. The Galaxy is estimated to be more than 30,000 parsecs in diameter.

Shapley's conclusion that the sun was not the center of the Galaxy came from his study of globular clusters. In 1917 he found that globular clusters are distributed in a roughly spherical shape that is not centered on the sun, but rather at a point some 10,000 parsecs or 32,000 light-years away (see Fig. 14.3). He also determined the diameter of this system of globular clusters and correctly assumed that this value also represented the size of the Milky Way Galaxy. His result was somewhat high; the presently accepted estimate of the Galaxy diameter is 30,000 parsecs or 100,000 light-years.

Like globular clusters, RR Lyrae stars have a roughly spherical distribution with respect to the center of the Galaxy. Therefore, this group of stars, too, can be employed to determine the distance to the center of the Galaxy. First the astronomer measures the number of RR Lyrae stars per unit volume as a function of the apparent magnitude in a line slightly off the galactic center as shown in Figure 14.4. Since the absolute magnitudes are the same for all of the RR Lyrae stars, a plot of apparent magnitude versus density can, by the use of the distance modululus formula, be readily converted into a plot of distance versus density. For a spherical distribution, the plot will pass through a maximum as shown in Fig. 14.4. The apparent magnitude at which the maximum occurs determines the distance r', and the distance r_0 to the center of the Galaxy is then easily obtained trigonometrically. Results from such analyses again yield a value of about 10,000 parsecs for r_0.

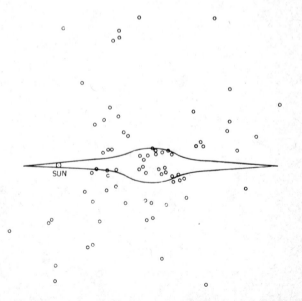

Fig. 14.3 A two-dimensional view of the distribution of the globular clusters. In three dimensions these objects form a roughly spherical distribution about the galactic nucleus.

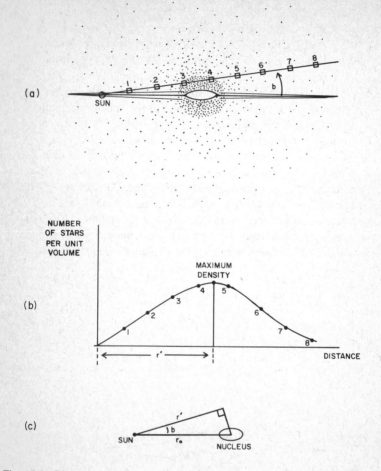

Fig. 14.4 Distance to the galactic center from counts of RR Lyrae stars.

MASS OF THE MILKY WAY

It is known that the sun moves about the center of the Milky Way in a roughly circular orbit having a radius of r_0. The velocity of the sun as it moves along this orbit can be computed from radial velocity measurements made on the globular clusters and the Magellanic Clouds (see p. 204), which are assumed to be at rest with respect to the center of the Milky Way. By dividing this orbital velocity (about 320 km/sec) into the total circumference of the sun's orbit, the sidereal period P_0 of the sun about the center of the Milky Way can be estimated and is found to be some 200 million years. Assuming that the mass of the Milky Way M_G is concentrated at the galactic center, it can be found by Kepler's harmonic law:

$$M_G = \frac{r_0{}^3}{P_0{}^2}$$

A simple calculation of M_G in solar masses yields a value of about 2×10^{11} solar masses, and since the average mass of a star is about one solar mass, this value also represents the approximate number of stars in the Milky Way Galaxy.

MAGNETIC FIELD OF THE MILKY WAY

Polarization studies of light from dust nebulae show an alignment of elongated particles due to the presence of a galactic magnetic field of about 10^{-6} gauss. This result is roughly confirmed by measurements made of Zeeman splitting in 21-cm radio emission from various parts of the Galaxy. The galactic magnetic field may be partly responsible for the acceleration of electrons and other atomic particles. These accelerated particles in turn give rise to the galactic radio noise as well as to much of the galactic cosmic ray flux.

STRUCTURE OF THE MILKY WAY

Our Galaxy can be thought of as consisting of three general, mutually interacting regions: the galactic plane or disk, the galactic nucleus, and the galactic halo or corona. Each of these regions exhibits a unique set of physical properties which are thought to offer important clues regarding the origin and evolution of the Milky Way.

The Galactic Plane. The plane of the Milky Way is not uniform in its content of stars and interstellar material, but exhibits a spiral structure not unlike that observed for the distant galaxies (see Chapter 15). The spiral arms contain the bulk of the stars, dust, and gas and hence can be traced out by mapping the locations of galactic star clusters, associations, and nebulosity as shown in Fig. 14.5. Unfortunately, because of the absorption effects in the visible region of the spectrum, it is not possible to map such objects beyond 1500 parsecs or so. As a

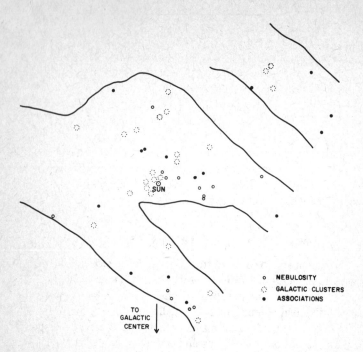

Fig. 14.5 Spiral structure in the solar vicinity.

result, the best mapping of galactic structure has been done by radio astronomers working at the longer wavelengths of radiation, which are essentially uninhibited by the interstellar medium. Of particular interest are the observations of the 21-cm line arising from hydrogen gas, the most abundant material in space. By observing the 21-cm line and measuring any radial velocities that are present, radio astronomers have been able to construct a picture of the Milky Way and have definitely established the existence of a spiral pattern in the distribution of hydrogen (see Fig. 14.6). Radio observations of other substances yield patterns of distribution that are similar to but not exactly the same as that found for hydrogen. The explanation of these differences is not known.

The establishment of a spiral structure in the Milky Way plane and the large solar motion relative to the center of the Milky Way strongly suggest that the entire Galaxy is rotating. If the average radial velocity of the stars is measured as a function of its galactic longitude or angular distance from the center of the Galaxy, the relative average velocity in all directions would be uniformly zero if the Milky Way were either not rotating or rotating as a rigid body. However, if the sun and the surrounding stars are moving in gravitationally generated Keplerian orbits, then the situation is that pictured in Fig. 14.7, where stars whose orbits are smaller than that of the sun are moving more rapidly than the sun and those with larger orbits are moving more slowly. From Fig. 14.7 it can be qualitatively argued that a plot of average radial velocity versus galactic longitude should result in a double sine curve, which is also depicted in Fig. 14.7. Observationally determined plots of mean radial velocity versus galactic longitude exhibit the double sine curve predicted for the Keplerian orbit of differential galactic rotation model and thus verify the existence of a galactic rotation. Of considerable interest in this regard is the discovery that whereas the outer regions of the Milky Way and other spiral galaxies are rotating as gravitating masses, the inner regions are rotating as rigid bodies with the rotational velocity increasing with increasing distance from the nucleus. For the Milky Way, the transition zone between rigid and nonrigid rotation occurs at 6000 or 7000 parsecs.

The Galactic Nucleus. At the center of the Milky Way spiral lies

Fig. 14.6 Spiral structure from radio observations of the 21-cm line. The sun's position is indicated by the arrow.

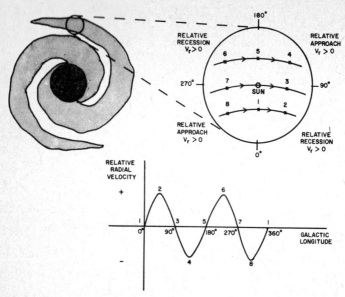

Fig. 14.7 Galactic rotation. If the mean radial velocity of stars in the solar vicinity is plotted against galactic longitude, the double sine curve (bottom) is the result.

the galactic nucleus, an oblate spheroidal concentration of stars about 4000 parsecs in diameter. Most of the visible light from the nucleus is blocked from the earth's view by interstellar gas and dust, and radio observations show that the nucleus is emitting long-wavelength non-thermal radio waves in a very complicated fashion. It is generally agreed that the center of the nucleus consists of a very dense assemblage of stars, possibly even a gigantic black hole, but the details of the structure of this region of the Galaxy are not known.

The Galactic Halo. Surrounding the Milky Way disk and nucleus is a vast spherical system of stars, gas, and globular clusters about 32,000 parsecs in diameter, which astronomers refer to as the galactic halo. Objects in the halo are believed to move in highly eccentric ellipses having the galactic nucleus at one focus. As a result of Kepler's law of equal areas, these objects move almost imperceptibly unless they are in the vicinity of the nucleus. Thus, to observers on the earth, who are part of the galactic rotation, these objects appear to have large radial velocities. Both nucleus and halo objects generally exhibit metal-poor spectra and are thus believed to occupy the oldest

Fig. 14.8 The satellite galaxies of the Milky Way Galaxy: (top) the Large Magellanic Cloud and (bottom) the Small Magellanic Cloud. (The Cerro Tololo Inter-American Observatory.)

parts of the Galaxy. As in the case of the nucleus, both thermal and nonthermal radio emissions have been detected from the halo, and like the nucleus, the structure and dynamics of this part of the Galaxy are not well understood.

THE MAGELLANIC CLOUDS

Like its sister galaxy in Andromeda, the Milky Way has two satellite galaxies that are visible in the Southern Hemisphere as the Large and Small Magellanic Clouds (see Fig. 14.8). Situated about 50,000 parsecs away, these objects have about one-tenth of the mass of the Milky Way and possess diameters of 8000 and 3000 parsecs, respectively. These galaxies have no rotational symmetry and are examples of a class of galaxies called *irregular galaxies*.

REVIEW QUESTIONS

1. Describe how the models for the size and shape of the Milky Way have evolved into our present view.
2. How is the size of the Milky Way Galaxy obtained? What is the location of the sun in this system?
3. If r_0 in astronomical units is 2×10^9 and the sidereal period of the sun is 200 million years, calculate the mass of the Milky Way from Kepler's harmonic law. *Ans.:* 2×10^{11} solar masses.
4. How is spiral structure deduced?
5. Sketch the Milky Way and indicate on your drawing the locations of (a) the spiral arms, (b) the galactic halo, (c) the galactic nucleus, and (d) the sun.
6. What are the Magellanic Clouds?

15

The Other Galaxies

By the start of the twentieth century all of the categories of deep-sky objects found in the New General Catalogue were understood except for the globular clusters and certain classes of diffuse objects that possessed a variety of symmetrical shapes, especially ellipses and pinwheels. Shapley, as we have seen in Chapters 13 and 14, demonstrated the spherical distribution of the globular clusters about the center of the Milky Way Galaxy and hence their intimate relationship with this stellar system. An explanation of the puzzling symmetrical nebulae, however, was not so easily obtained. Unlike the globular clusters, which possessed spherical symmetry relative to the galactic nucleus, and the gaseous nebulae and open star clusters, which are found almost exclusively along the Milky Way plane, the symmetrical nebulae seemed to be collected at the galactic poles. In fact, these objects were found to be absent from a region called the *zone of avoidance,* which closely coincides with the visible regions of the Milky Way. Some astronomers assumed that these objects were part of the Milky Way, whereas others thought of them as "island universes," systems of stars comparable to the Milky Way in size but millions of light-years away.

The debate raged for nearly three decades before the great American astronomer Edwin Hubble resolved the controversy in 1924 in favor of the island-universe theory. Using the newly completed 100-inch reflecting telescope at the Mount Wilson Observatory, Hubble discovered cepheid variables in the outer regions of the Andromeda and Triangulum "nebulae" and found that their apparent brightnesses were extremely low despite their large intrinsic brightnesses. He concluded that these nebulous patches were very remote and well beyond the boundaries of the Milky Way. The galaxies, as we now call them, came to be recognized as separate celestial entities, systems of stars whose dimensions and stellar content rival those of the Milky Way Galaxy.

DISTRIBUTION OF GALAXIES

Galaxies, like stars, are found throughout the sky, but only a few, such as the Andromeda galaxy and the Magellanic Clouds, are bright enough to be seen with the naked eye or a small-aperture telescope. At fainter magnitudes the total number of detectable galaxies increases dramatically.

The Zone of Avoidance. It was early noticed that galaxies are almost completely absent from a band of the sky centered on the plane of the Milky Way. This "zone of avoidance" is caused by clouds of obscuring material that block our view to the more distant galaxies. As the astronomer looks outward from the plane of the Milky Way, more and more galaxies per square degree are observed because there is less and less obscuring material. One would expect, then, that the highest number of galaxies per square degree would be observed in a direction perpendicular to the plane of the Milky Way where the amount of obscuring material is at a minimum; this is indeed the case.

Clusters of Galaxies. Galaxies tend to occur in clusters (see Fig. 15.1). The membership of clusters can range from only a few galaxies up to several thousand. The Milky Way, the Magellanic Clouds, and the Andromeda galaxy all belong to one such cluster, which astronomers refer to as the *Local Group*. Some of these clusters, usually those having large numbers of galaxies, have spherical symmetry and central concentration; they are called *regular clusters* or *globular clusters*. Other clusters, such as the Hercules cluster shown in Fig. 15.2, possess little or no spherical symmetry or central concentration and are referred to as *irregular clusters* or *open clusters* (see Table 15.1).

The dimensions of these clusters of galaxies are enormous; for example, it is estimated that the linear diameter of the Coma cluster of

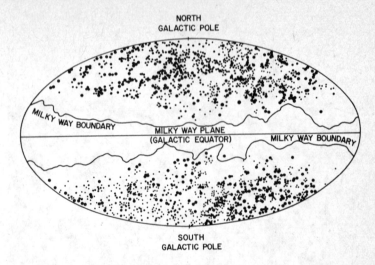

Fig. 15.1 An idealized distribution of clusters of galaxies on the celestial sphere. The size of each cluster is roughly proportional to the size of the dot. The dark, irregular lines are the approximate boundaries of the Milky Way.

galaxies is of the order of 5 million parsecs and that the mean distance between galaxies is of the order of several thousand parsecs.

Astronomers have also found that the clusters of galaxies themselves are not uniformly distributed but tend to be grouped into *second-order clusters, clusters of clusters,* or *superclusters.* These

Table 15.1. General Properties of Selected Clusters of Galaxies

Cluster	Number of Galaxies	Angular Diameter (degrees)	Distance (parsecs)
Local Group	17	—	4.0×10^5
Virgo	2500	12	1.1×10^7
Perseus	500	4	5.8×10^7
Coma	1000	6	6.8×10^7
Hercules	75	1	1.1×10^8
Corona Borealis	400	0.5	1.9×10^8
Bootes	150	0.3	3.8×10^8
Ursa Major	200	0.2	3.8×10^8

Fig. 15.2 The Hercules cluster of galaxies. Nearly every image on this photograph is a galactic system comparable in size to the Milky Way. (Courtesy of the Hale Observatories.)

systems may have diameters of between 30 and 50 million parsecs and total masses of 10^{15} solar masses.

The possibility of *third-order clustering* of galaxies has also been considered, but the existence of such systems has yet to be demonstrated to the satisfaction of all astronomers.

CLASSES OF GALAXIES

Galaxies are usually grouped into four general classes: the ellipticals, the normal spirals, the barred spirals, and the irregulars. Of the several detailed calssification schemes for galaxies that have been proposed, the simplest is that suggested by Edwin Hubble (see Fig. 15.3).

Ellipticals. In the Hubble system, elliptical galaxies are classified according to their degree of ellipticity, or flattening, which is defined as $10(a-b)/a$, where a and b are the major and minor axes of the image of the galaxy. The numbers designating the degree of ellipticity range from 0 to 7. A capital letter E is placed in front of the degree of

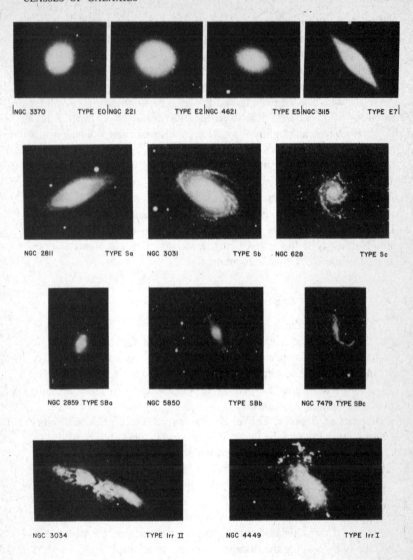

NGC 3370 TYPE E0 NGC 221 TYPE E2 NGC 4621 TYPE E5 NGC 3115 TYPE E7

NGC 2811 TYPE Sa NGC 3031 TYPE Sb NGC 628 TYPE Sc

NGC 2859 TYPE SBa NGC 5850 TYPE SBb NGC 7479 TYPE SBc

NGC 3034 TYPE Irr II NGC 4449 TYPE Irr I

Fig. 15.3 The basic Hubble classes of galaxies. (Courtesy of the Hale Observatories.)

flattening to denote an elliptical galaxy. Thus an elliptical galaxy having a spherical shape would be classified as E0; an extremely flat one would be designed E7.

Spirals. Normal spirals are designated by S and barred spirals by SB. To these designations the letters a, b, and c are added to denote the relative size of the nucleus and the degree of tightness to which the spiral arms are wound. Thus type Sa indicates a galaxy with a large nucleus, no bar, and tightly wound arms, whereas type Sc is the designation given to a galaxy with no bar, a small nucleus, and loosely wound arms. Galaxies that are intermediate between Sa and Sc are classified as type Sb. If a bar is present, the corresponding subtypes are SBa, SBb, and SBc.

Irregulars. Irregular galaxies are classified as either Irr I or Irr II. The Irr II galaxies show prominent dark nebulosity; Irr I galaxies do not. Neither type has a trace of the rotational or circular symmetry present in the ellipticals and spirals.

DISTANCES TO GALAXIES

The distance to the galaxies, even to the nearby systems, are so vast that the methods of trigonometric parallax used for determining fundamental stellar distances are totally inadequate. As a result, to obtain distances to galaxies, astronomers must resort to indirect methods that are accurate to only a few percent at best and in many cases are only order-of-magnitude estimates.

Distance Indicators. The most reliable methods of galactic distance determination involve the use of distance indicators, objects whose intrinsic brightnesses are known and which have been identified in other galaxies. Some of the objects used by astronomers in this regard are listed in Table 15.2 and are illustrated in Fig. 15.4. By comparing the known absolute magnitude of a given distance indicator with the measured apparent magnitude of the same type of object that has been located in another galaxy, astronomers can determine the distance to the galaxy by solving the distance modulus relationship

$$m - M = 5 \log r - 5 \qquad \text{for the distance } r$$

For example, RR Lyrae stars have been detected in the Andromeda galaxy, and the apparent magnitude of these objects has been found to be $+24$. Since the absolute magnitude of RR Lyrae stars is known to be 0.0, the distance modulus $m - M$ is 24, and the corresponding dis-

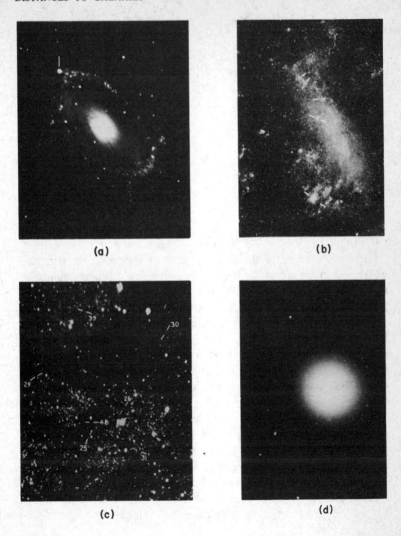

(a)

(b)

(c)

(d)

Fig. 15.4 Examples of galactic distance indicators: (a) a supernova in NGC 4725, (b) H II regions in the Large Magellanic Cloud, (c) Cepheid variables in the Andromeda galaxy M31, and (d) globular clusters about the fringes of M87 in Virgo. By comparing the apparent brightnesses of such objects with their known intrinsic brightnesses, the distances to galaxies can be estimated. [(a), (c) Courtesy of the Hale Observatories; (b) by permission of the Harvard College Observatory, Cambridge, Mass.; (d) Lick Observatory photographs.]

Table 15.2.   Galactic Distance Indicators

Object	Approximate Absolute Magnitude	Approximate Limiting Distance of Detection (parsecs)
Cepheid variables	− 6	7×10^6
RR Lyrae stars	0	3×10^5
H II regions	− 9	3×10^7
Red giants and supergiants	− 3	2×10^6
O and B supergiants	− 9	3×10^7
Globular clusters	− 9	3×10^7
Novae	− 9	3×10^7
Supernovae	−19	3×10^9

tance to M31 is about 680,000 parsecs. This method of distance determination is essentially the same as that of spectroscopic parallax, which was discussed in Chapter 10. Of course, the basic assumption in such an analysis is that the physical properties, in particular the absolute magnitudes, of the objects employed as distance indicators do not vary appreciably from galaxy to galaxy. No evidence to the contrary has yet been found.

Determination of Radial Velocity. In 1929 Edwin Hubble found that the radial velocity V_r of a galaxy measured from the Doppler shift of the lines in its spectrum was directly proportional to the distance of the galaxy, or $V_r = Hr$, where H is the Hubble constant. (See Fig. 15.5.) If there are no observable distance indicators in a given galaxy, it is thus still theoretically possible to obtain an estimate of the distance to the galaxy from a determination of V_r. Unfortunately, the procedure requires that the value of H be accurately known, and measurements of the value of H range from 50 to 150 km/sec/million parsecs, or differ by a factor of 3. Some astronomers have suggested that the value of H be defined as 100 km/sec/million parsecs.

LINEAR DIMENSIONS OF GALAXIES

Once the distance r to a galaxy has been determined, it is possible to calculate its linear dimaeter d from the apparent angular diameter α (in arcseconds) = 206,265 d/r. Angular diameters of galaxies are difficult to determine accurately owing to the diffuse nature of these objects, but based on the currently accepted values for galactic angular

RELATION BETWEEN RED SHIFT AND DISTANCE FOR EXTERIOR GALAXIES

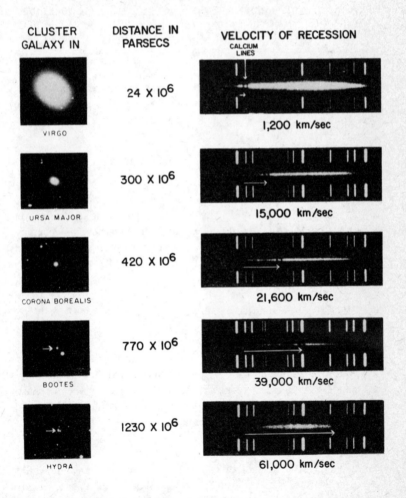

CLUSTER GALAXY IN	DISTANCE IN PARSECS	VELOCITY OF RECESSION
VIRGO	24×10^6	1,200 km/sec
URSA MAJOR	300×10^6	15,000 km/sec
CORONA BOREALIS	420×10^6	21,600 km/sec
BOOTES	770×10^6	39,000 km/sec
HYDRA	1230×10^6	61,000 km/sec

Fig. 15.5 The velocity-distance relation for galaxies. As the distance to a galaxy increases, the observed radial velocity, as measured from the red shift of its spectral lines, also increases. (Courtesy of the Hale Observatories.)

Table 15.3. Linear Dimensions and Luminosities of Galaxies

Type	Linear Diameter (parsecs)	Absolute Magnitude	Luminosities ($L/L_\odot$)
Ellipticals	700 to 2×10^5	-10 to -23	10^6 to 10^{11}
Spirals	7×10^3 to 5×10^4	-15 to -20	10^8 to 10^{10}
Irregulars	3×10^3 to 10^4	-13 to -18	10^7 to 10^9

diameters, the corresponding linear diameters range from a few hundred parsecs for the dwarf ellipticals up to 50,000 parsecs for the large spirals and ellipticals. (See Table 15.3.)

LUMINOSITIES OF GALAXIES

If the apparent magnitude of a galaxy is known, the absolute magnitude can be obtained from the distance modulus formula. Again the diffuse, extended nature of a galaxy makes an accurate determination of an apparent magnitude difficult, which in turn restricts the accuracy of the absolute magnitude. Absolute magnitudes of galaxies range from -1 to -23, which makes galaxies between 10^6 and 10^{11} times as luminous as the sun.

SPECTRA OF GALAXIES

Because they are vast systems made up of individual stars, the galaxies display composite absorption line spectra that range from type A to K for the spirals, G to K for the ellipticals, and A to F for the irregulars. The ellipticals are generally of later spectral types because they are made up almost entirely of late-type Population II stars whereas the spirals and irregulars contain a mixture of both early- and late-type stars.

Some peculiar galaxies such as the Seyfert galaxies display spectra that contain broad, bright emission lines. These objects are often strong radio emitters as well, and are not well understood at present. It is felt, however, that Seyfert galaxies may be closely related to quasars (see page 219) and the nuclei of normal galaxies.

The fact that all the exterior galaxies exhibit a redshift in their spectral lines is thought to be an indication that these objects are receding from the earth. The velocity of recession also increases with galactic distance as shown in Fig. 15.5, the constant of proportionality being the previously mentioned Hubble constant.

MASSES OF GALAXIES

The masses of galaxies are obtained by measuring their gravitational effects on other galaxies or on individual stars and objects within a given galaxy. If a galaxy is relatively near, spectroscopic means can be used to estimate the orbital velocities of stars in its various regions. If the distance to the galaxy is known and some sort of mass distribution is assumed, then the mass of the galaxy can be computed.

Let us consider the edge-on galaxy shown in Fig. 15.6. By measur-

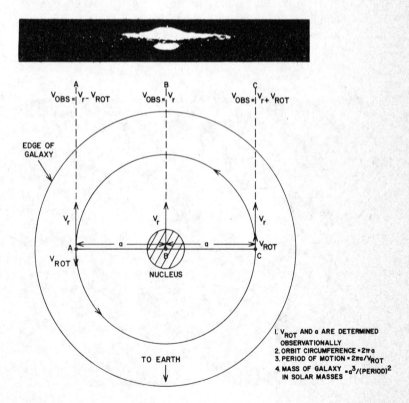

Fig. 15.6 Determining the mass of an edge-on galaxy. The bulk of the galactic mass is assumed to be at point B. (Courtesy of the Hale Observatories.)

ing the Doppler shift at points A and B, it is possible to determine the overall radial velocity of the galaxy, V_r, as well as the rotational velocity, V_{rot}, at point A. If the distance to the galaxy is known or can be determined, then the linear distance a can be obtained from the measured angular distance between points A and B. We now assume that the stars at point A are moving in circular orbits on the average, and that V_{rot} represents the average orbital velocity of these stars. The period P of the motion is then equal to the orbit circumference at A divided by the rotational velocity at A, or $P = 2\pi a/V_{rot}$. By making the simple assumption that the mass of the galaxy is concentrated at the center, Kepler's third law, $\mathbf{M}_g = a^3/p^2$, may be used to calculate the mass of the galaxy, $\mathbf{M}_g$. This calculation is very similar to that performed for the Milky Way Galaxy in Chapter 14. This is, of course, a highly simplified version of the problem. In practice, the astronomer must consider such factors as the galaxy's space orientation and mass distribution.

Masses may also be calculated from observed motions of binary

Fig. 15.7 The binary galaxy NGC 5432 and NGC 5435. By studying the motions of large numbers of such systems, astronomers can deduce the average masses of galaxies. (Lick Observatory photograph.)

galaxies and galaxies within clusters (see Fig. 15.7). The methods employed in these cases are statistical in nature and yield only average masses.

Masses of galaxies calculated from the above techniques are found to range from 10^7 solar masses for the tiny dwarf systems to 10^{13} solar masses for some of the giant elliptical galaxies.

EVOLUTION OF GALAXIES

The various morphological forms of galaxies can be arranged in a nearly continuous sequence of structural types starting with the ellipticals and ending with the irregulars as shown in Fig. 15.8. This fact immediately suggests that the different types of galaxies represent various stages of evolution. According to one theory, galaxies are born as E0 ellipticals, then flatten, develop spiral arms, and evolve into the spiral and barred spiral galaxies. There are several difficulties with this theory, and most modern astronomers who believe that galaxies do evolve suggest a reverse evolutionary sequence based on the gas, dust, and stellar population types present in the galaxies. In these latter theories, galaxies are thought to begin as irregulars, evolve through the various stages of spirals, and end up as ellipticals. There is considerable doubt, however, that the Hubble types represent any kind of evolutionary sequence at all; in fact, it may be that all the galaxies have retained their present structures from the time of their formation.

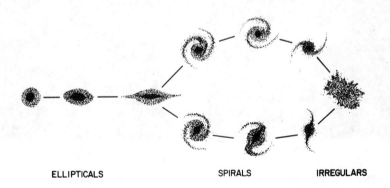

ELLIPTICALS   SPIRALS   IRREGULARS

Fig. 15.8 Sequence of Hubble types of galaxies. The smooth transition from one type of galaxy to another has led some astronomers to suspect the existence of some sort of galactic evolutionary scheme.

RADIO GALAXIES

Since galaxies are made up of large numbers of stars, the continuous spectrum of these objects is roughly comparable to that of a blackbody; thus only a small fraction of the energy output of the *normal* galaxies occurs in the radio region of the spectrum. Some galaxies, however, emit unusually large amounts of radio energy that can be as much as a million times greater than what would be expected. Many of these *peculiar* radio galaxies also display unusual visual characteristics that seem to be associated with their large amounts of radio emission (see Fig. 15.9). For example, Centaurus A, a strong radio

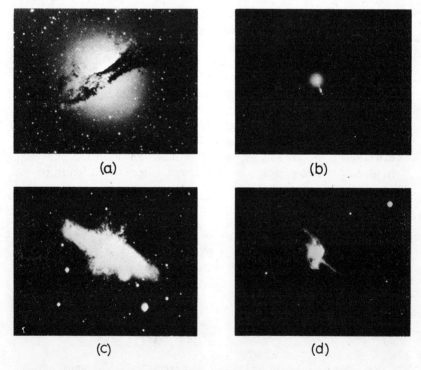

Fig. 15.9 Four peculiar radio sources: (a) NGC 5128 (Centaurus A), (b) the nucleus of M82, (c) the explosion in the core of M82, and (d) the colliding system NGC 2623. The relationship between the observed peculiarities and the large radio emissions from these objects is not well understood. [(a), (c), (d) Courtesy of the Hale Observatories; (b) Lick Observatory photograph.]

source in Centaurus, shows up on photographs as an elliptical galaxy that possesses a broad dust lane. A number of peculiar radio galaxies display jets of material protruding from their nuclei, whereas others show up on photographic plates as pairs of galaxies in apparent collision.

These radio sources are extremely complex, and the theories dealing with the mechanism for this output of energy have not generally been satisfactory. One of the more reasonable theories is that the radio emissions are a form of synchrotron radiation in which the observed radio energy is generated by the acceleration of electrons in a magnetic field. Other frequently mentioned mechanisms include chain reactions between supernovae, fission of galactic nuclei, and conversion of galactic gravitational potential energy into the required radio energy by means of galactic gravitational collapse.

QUASARS

During the course of the investigation of radio sources in the late 1950s, astronomers discovered several sources of radio emission that, unlike previously observed sources, exhibited star-like images when photographed at visible wavelengths. Spectra were taken of these objects and it was found that the lines present were shifted far forward from their normal wavelengths. If these large redshifts were interpreted as Doppler shifts, then these objects must be moving away from the earth at enormous velocities. Comparatively nearby bodies moving at such velocities would be expected to display some detectable proper motion, but since none was observed for these objects, it was concluded that their distances were well beyond the boundaries of the Milky Way. Their star-like appearance was therefore not regarded as intrinsic but due rather to the inability of telescopes to resolve any appreciable structure at the distances involved. These newly discovered objects were thus called *quasi-stellar radio sources* (QSSs) or quasars.

General Characteristics. Since 1960 well over 100 quasars have been discovered. They are luminous objects, stellar in appearance when photographed in visible light; a few, such as 3C 273 (Fig. 15.10), have tiny whisps of jets of material emanating from their nuclei. Quasars are blue in color and radiate heavily in the ultraviolet. Their radio emission and visible light are variable, with fluctuations of as much as one magnitude. The spectra of quasars can contain both

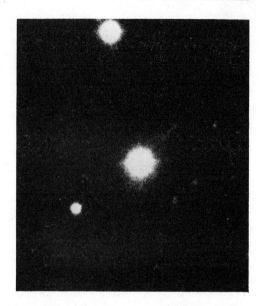

Fig. 15.10 The quasi-stellar radio source 3C 273. Note the faint jet of material issuing from the nucleus. (The Kitt Peak National Observatory.)

emission and absorption lines, but they are generally of the emission-line type. The lines present are usually highly redshifted, most quasars having $\Delta\lambda/\lambda$ values between 1.0 and 2.3. (See Fig. 15.11.) There are also radio-quiet quasars called *quasi-stellar galaxies* (QSGs), which display all of the quasar characteristics except that of high radio emission.

Theories of Quasars. It is almost certain that quasars are extragalactic objects, but beyond this little is known of their true nature and they remain perhaps the most puzzling objects in all of astronomy. Most of the controversy regarding quasars centers on the interpretation of the observed redshifts. The most commonly accepted idea is that the redshift is both Doppler and *cosmological*, that is, the redshift represents velocity of recession as well as a direct indicator of distance. If this interpretation is accepted and the other observed properties of quasars are taken into account, more than 100 times as much energy as is produced by all the stars in the Milky Way must be released within a region whose volume is 10^{-17} that of the Mily Way. The basic problem thus becomes one of finding an energy-generating mechanism that satisfies all of the observed data, and no mechanism thus far proposed does this satisfactorily.

A second approach is to assume that the redshifts of the quasars are

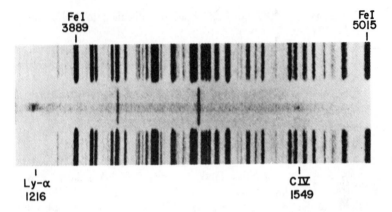

Fig. 15.11 The spectrum of the quasar PKS 0229 +13. In this object, the emission lines of Lyman-α and C IV 1549 (center spectrum) are shifted over 1500 Å toward the red! The bright, sharp emission lines in the center spectrum are from city lights. Iron comparison lines are above and below the quasar spectrum. (The Kitt Peak National Observatory.)

Doppler shifts but are not cosmological. Thus a quasar could be much closer to the Milky Way than a galaxy having the same radial velocity. This idea eases the astronomer out of the problem of finding a suitable energy-generating mechanism, but leaves the unanswered questions of the origin of such objects and the source of their enormous velocities.

Other astronomers suggest that the observed redshifts are not Doppler shifts at all, but arise from some other phenomenon in nature. Aside from redshifts produced by large gravitational fields, no other redshift mechanisms are known to exist in physics either theoretically or experimentally. Unfortunately, no configuration of matter thus far proposed is able to fit the observed characteristics of the quasars and still produce the required redshifts gravitationally.

There are many other theories designed to explain quasars, but to date none completely accounts for all of their observed properties.

REVIEW QUESTIONS

1. Explain how the zone of avoidance can be accounted for in the island-universe theory of galaxies.
2. Describe the Hubble classification scheme for galaxies.
3. The average apparent magnitude of the globular clusters observed

in the galaxy M87 is $+22$. Find the distance to M87. *Ans.:* 1.6×10^7 pc.

4. A galaxy is observed to have a radial velocity of 60,000 km/sec. If **H** is equal to 100 km/sec/million pc, estimate the distance to the galaxy. *Ans.:* 600 million pc.

5. Describe how you would determine the linear size of a galaxy. What information would you need to know about the galaxy in order to obtain its diameter?

6. An edge-on galaxy has an observed radial velocity of $+1000$ km/sec at the nucleus and $+1200$ km/sec at a point 10,000 parsecs from the nucelus. Find the mass of this galaxy. *Ans.:* 8.1×10^{10} solar masses.

7. Describe the spectrum of a galaxy. How does it compare with the spectrum of a star?

8. What conclusion may we draw from the fact that all galaxies exhibit redshifts in their spectra?

9. Discuss the evidence for and against the evolution of galaxies.

10. Why are quasars puzzling to astronomers?

11. What is the difference between a galactic cluster and a cluster of galaxies?

16

Cosmology

Despite the vastness of space and the diversity of objects and star systems that have been observed, astronomers try to attain an overview of the universe, one that describes its nature and evolution. The pursuit of such an overview is referred to as cosmology. In a matter of only a few generations, our view of the universe has changed dramatically. The Renaissance struggle between the heliocentric and geocentric models for the planetary system took the earth from the center of the universe to a small planet orbiting the sun. The debate early in this century concerning the nature of the globular clusters and the diffuse galaxies has led to the discovery that the sun is on the outskirts of the vast Milky Way system of some 200 billion suns, which itself is only one of millions of such galaxies known to exist in the observable universe. Such is the staggering view of the universe presented by cosmologists in the middle of the twentieth century.

GENERAL PROPERTIES

In arriving at a modern cosmology, it is first necessary to consider the overall characteristics of the observable universe.

The Expansion of the Universe. As noted in the last chapter, every one of the distant galaxies outside of the Local Group exhibits a redshift in the positions of its spectral lines. This redshift, first observed by V. M. Slipher, is interpreted by astronomers as a Doppler shift arising from the velocities of recession of these objects. Thus, the first property that can be assigned to the universe as a whole is that it appears to be expanding, with every galaxy receding from every other galaxy much as the spots on a balloon recede from one another as the balloon is blown up.

Further studies by Hubble established that the velocity of recession increases linearly with distance where the constant of proportionality is the Hubble constant H (see p. 212). It is reasonably certain that the linearity of the Hubble law is preserved out to several hundred million parsecs, but beyond this distance, the uncertainties in the determined distances to galaxies are such that any kind of subtle departure from linearity brought about by a change in the value of H or a change in the basic nature of the Hubble law itself would go virtually undetected.

"Fossil" Radiation. A crucial discovery regarding the overall properties of the universe was provided in 1965 by Penzias and Wilson of the Bell Telephone Laboratories when they announced the existence of very weak radio emission at a wavelength of 7.4 cm that was isotropic, that is, equally intense in all directions of space. By 1970, similar radiation was observed at five additional radio wavelengths ranging from 0.86 to 20 cm. Moreover, careful studies of the lines arising from interstellar cyanogen (CN) molecules indicated a level of thermal excitation arising from a background source of interstellar radiation. All of these observations can be accounted for by assuming the existence of a general level of background radiation in space that has the capability of heating a blackbody to a temperature of some 2.8°K (see Fig. 16.1). This radiation is generally believed to be "fossil" radiation or the radiation that was left over from a primeval explosion that started the presently observed expansion of the universe.

The Density of the Universe. The mean density of the matter present in the universe can be readily calculated from the ratio of the total mass of the universe to its total volume. This computation is usually accomplished by conducting galaxy counts as a function of distance and assigning a mean mass value to each observed galaxy which is obtained from mass determinations made for more accessible galaxies. Because of the large uncertainties in both the distances and the masses of galaxies, the resulting, value for the mean density of the universe is highly uncertain, and ranges from 10^{-29} to 10^{-31} g/cm^3.

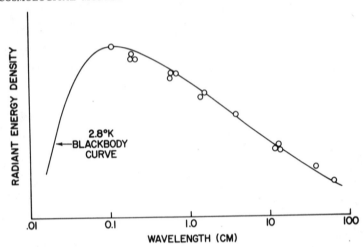

Fig. 16.1 Observations of the cosmic microwave background radiation. The solid curve is the theoretical energy distribution by wavelength for a 2.8°K blackbody.

COSMOLOGICAL MODELS

Astronomers have developed three basic cosmological models, the Big Bang theory, the oscillating theory, and the steady-state theory. (See Fig. 16.2.)

The Big Bang Theory. The Big Bang theory was first proposed in the 1920s by the Belgian cosmologist Abbe Lemaitre. According to this theory, the universe began as a "primeval atom," a chunk of matter roughly the size of the solar system that exploded about 15 billion years ago into an enormous fireball with a temperature of trillions of degrees. From this initial "big bang," the material in the universe was hurled in all directions, and continues to this day to expand. As the mean temperature of the fireball decreased, the expanding material subsequently cooled and contracted into the galaxies. The Big Bang theory implies that the universe had a definite beginning and therefore has a definite lifetime.

The Oscillating Theory. In the oscillating theory, material is hurled outward from a Big Bang-type fireball, but instead of the matter expanding outward indefinitely, it slows gradually because of internal self-gravitation until it stops and then begins to collapse. The material then increases in density until the conditions of the fireball are repro-

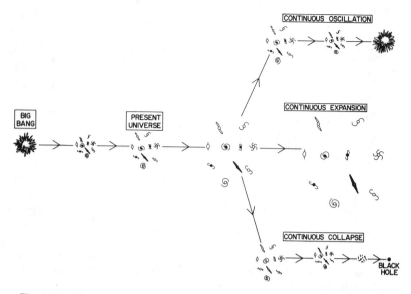

Fig. 16.2 The basic cosmological models: the single Big Bang model that forever expands, the multiple Big Bang or oscillating model, and the single Big Bang model that expands outward for a while and then forever collapses from self-gravity.

duced and the expansion begins anew. The oscillating theory thus views the universe as an eternally pulsating entity that completes a single pulsation cycle about once every 40 to 50 billion years. At the present time, of course, the universe is on the expanding part of the cycle.

One recent variation of this theory hypothesizes that after the outward expansion ceases, the material in the universe will contract and finally collapse into a gigantic black hole. Thus after the first "big bang" there is only a single expansion, followed by a single contraction.

The Steady-State Theory. In 1948, the English cosmologists T. Gold, H. Bondi, and F. Hoyle proposed yet another model for the universe called the steady-state theory. This theory postulates that not only is the existence of matter in the universe eternal, but also the mean density of matter in the universe remains constant in time. As the universe expands, clearly the density must decrease unless matter somehow is continuously introduced into the "gaps" left by the expanding material. To this end, advocates of the steady-state theory have postulated the continuous creation of matter, or the appearance of matter from out of the void at just the right rate (about 1 proton/cubic km/year) to maintain a constant density. Thus the universe that is visible to us now has exactly the same observable properties that it has had at any time in the past or will have at any time in the future. Although such a picture of continuous creation might seem a strange way of viewing things, it is philosophically no more elegant to assume that the material of the universe has suddenly come into being from out of the void than to assume continuous creation from out of the void.

TESTS OF THE COSMOLOGICAL MODELS

Each of the cosmological models discussed above predicts a unique set of observable characteristics for the state of the universe at the present time as well as for the dynamics of its past history. In the latter case, the vast distances that plague the level of accuracy attainable by the observational astronomers also permit them to look back into time. For example, if the distance to an object is 1 million light-years, the light that astronomers view from that object left there 1 million years ago, and hence portrays that object as it existed at that time. Thus, by observing objects at ever greater distances, astronomers can view the state of the universe at increasingly earlier times in its history. In this fashion, astronomers are able to infer the past history of the universe, but only within the framewok of observational data that at present are laden with considerable uncertainties.

The Expansion of the Universe. Although all of the cosmological models predict the observed expansion of the universe, the rate of the expansion in time differs with each model. The steady-state theory requires a uniform expansion not only in space but also in time. The oscillating and Big Bang models, on the other hand, predict that the rate of expansion decreases with time. The oscillating universe, however, requires a significantly higher slowdown rate than does the Big Bang, since the outward expansion of the oscillating universe must, at some point in time, ultimately stop as gravitational contraction sets in.

Measurements of the rate of expansion of the universe made on galaxies several billion light-years away indicate a higher expansion rate for the universe at the time the light left these objects than that which is observed for closer galaxies whose light-travel time is much less. Thus, despite the uncertainties present in the measurements, data

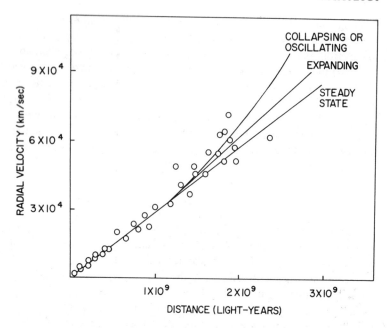

Fig. 16.3 A cosmological test involving the Hubble law. Each of the basic cosmologies predicts a different behavior for the distance-radial velocity curve. The points plotted represent the scatter of the present observations.

obtained to date on the expansion of the universe suggest a universe whose expansion rate is decreasing in time, thus excluding the steady-state theory. The data, however, are not sufficiently accurate to eliminate definitely either the oscillating or Big Bang theories (see Fig. 16.3).

Fossil Radiation. The discovery of fossil radiation in space with a blackbody temperature of about 3°K was a devastating blow to the steady-state theory, since that theory cannot in any fashion account for such an observation. On the other hand, the existence of such radiation is a logical consequence of the primeval fireball of the Big Bang and oscillating theories. The intense radiative energy that must have accompanied such an event has apparently now spread out over an ever-expanding universe to the extent that only a very low energy density of fossil radiation remains. No test has been devised to eliminate either the Big Bang or the oscillating theory on the basis of the predicted properties of the fossil radiation.

The Density of the Universe. The various cosmological models also make differing predictions concerning both the distribution of material in the universe and its behavior in time, and as a result, these properties too can serve as discriminators of cosmological models.

As one looks to larger distances and hence backward in time, the density of material usually obtained from the number of galaxies observed per unit volume ought to increase for both the oscillating and the Big Bang models, the higher increase being predicted for the oscillating universe. The density of the steady-state universe, of course, remains constant. Once more, however, a possible cosmological test suffers from a lack of accuracy in the data currently available. An additional problem may also arise in that there is increasing evidence that a significant amount of material in the universe may be present in a virtually undetectable form, such as black dwarfs, neutron stars, and intergalactic matter.

Another point of interest is the fact that if the universe has a mean density of less than 10^{-29} g/cm^3, the attractive gravitational forces will not be large enough to stop the presently observed rate of expansion of the universe. Thus, depending on the value found for the mean density of the universe, it is possible to distinguish by observational means between the oscillating theory, which must provide the density needed ultimately to stop the expansion, and the forever-expanding steady-state and Big Bang models. Estimates of the density of the material present in the universe, as noted above, range from 10^{-29} to 10^{-31} g/cm^3, a result that suggests there may not be enough material in the universe for its outward expansion to be stopped by gravitational forces. With an uncertainty of two orders of magnitude, however, this result is far from definitive.

GEOMETRY OF THE UNIVERSE

Until the twentieth century, scientists had assured themselves that the straight-line geometry developed centuries ago by Euclid fully described all spatial relationships in the universe. In 1916, however, in his theory of general relativity, Albert Einstein postulated the idea that light will always travel along a *geodesic,* the shortest distance between two points in a given space. Since the given space is mathematically defined in terms of its geodesic, Einstein's contention effectively eliminates the need for forces by stating that the space is defined by the path in which light is observed to travel. In this four-dimensional

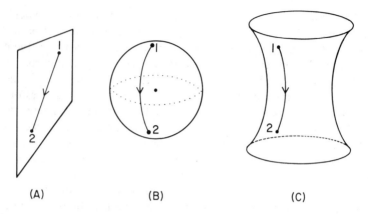

Fig. 16.4 Two-dimensional spaces with (a) zero curvature, (b) positive curvature, and (c) negative curvature. The line from point 1 to point 2 represents the shortest distance between the points of the geodesic for that space.

space-time, the paths of light ray can be bent or deflected by the presence of massive objects. Thus, one of the consequences of general relativity is that the presence of matter in the universe imparts a general curvature or warp to space that may be positive (convex), negative (concave), or zero, depending on the mean density of the universe. Such a curvature in Einstein's four-dimensional space is rather difficult to visualize, but some idea of the principles involved can be gained by considering the equivalent two-dimensional spaces shown in Fig. 16.4 along with their respective geodesics. Unfortunately, the same uncertainty in the density of the universe that defeated the classical force description of its dynamics prevents the relativistic description from ascertaining the effect of the material in the universe on the light geodesic. Hence the geometry of space has thus far eluded the best efforts of relativistic cosmologists.

OTHER UNIVERSES

Each jump in our perception of the size of the universe has brought with it the discovery that other entities exist that possess the properties of the erstwhile universe. Thus, the solar system, whose boundaries were for so long thought to mark the extent of the universe, almost certainly has billions of counterparts among the stars of the Milky

Way and other galaxies. Much the same can be said of the Milky Way itself. Herschel, Kapteyn, Shapley, and others had pictured the Milky Way as the sum total of the universe, and this view held for two decades in this century, until it was discovered that the Milky Way itself is but one of millions of such systems strewn across space. Moreover, both the Big Bang and oscillating cosmological models implicitly suggest that the size of the universe is finite. It is only natural, therefore, that cosmologists have speculated on the possible existence of other "universes." This question, unfortunately, cannot be answered beyond a speculative level with the instrumentation currently available. Perhaps the gigantic orbiting telescopes scheduled for launch in the last decades of this century, which will have the capability of peering more deeply into space with much greater resolution, will not only shed light on this question, but may also serve to reduce greatly the magnitude of Hubble's "ghostly errors of measurement" that have haunted cosmologists throughout astronomical history.

REVIEW QUESTIONS

1. Describe some of the observable properties of the universe.
2. Compare and contrast the Big Bang, oscillating, and steady-state theories of the universe.
3. What predictions are made by the three cosmological models regarding the behavior the observable properties of the universe?
4. To what extent does each theory successfully account for the observations? Explain.

Appendices

IMPORTANT PHYSICAL AND ASTRONOMICAL CONSTANTS

Astronomical unit (AU)	1.496×10^{13} cm
	9.296×10^{6} mi
Light-year (Ly)	9.461×10^{17} cm
	6.324×10^{4} AU
	5.879×10^{12} mi
Parsec (pc)	2.063×10^{5} AU
	3.262 Ly
	3.086×10^{18} cm
	1.808×10^{13} mi
Mass of earth ($M_{\oplus}$)	5.977×10^{27} grams
Mass of sun ($M_{\odot}$)	1.991×10^{33} grams
Luminosity of sun ($L_{\odot}$)	3.862×10^{33} ergs/sec
Radius of sun ($R_{\odot}$)	6.960×10^{10} cm
	4.322×10^{5} mi
Velocity of light (c)	2.998×10^{10} cm/sec
	1.862×10^{5} mi/sec
Universal gravitation constant (G)	6.668×10^{-8} dyne-cm²/g²
Stefan-Boltzmann constant (σ)	5.669×10^{-5} erg/cm²-°K⁴-sec
Wien's law constant (W)	2.898×10^{-1} cm-°K
Planck's constant (h)	6.624×10^{-27} erg-sec
Boltzmann's constant (k)	1.380×10^{-16} erg/°K

APPENDIX 2
PROPERTIES OF THE PLANETS

Planet	Mean Diameter (km)	Mass ($M_\oplus$)	Albedo	Rotation Period	Mean Radius of Orbit (AU)	Sidereal Period (years)	Orbital Inclination (degrees)	Orbital Eccentricity
Mercury	4,890	0.06	0.06	59^d	0.39	0.24	7.0	0.21
Venus	12,100	0.82	0.76	243^d	0.72	0.62	3.4	0.01
Earth	12,740	1.00	0.39	24^h00^m	1.00	1.00	0.0	0.02
Mars	6,760	0.11	0.15	24^h37^m	1.52	1.88	1.9	0.09
Jupiter	143,200	318.0	0.51	9^h50^m	5.20	11.9	1.3	0.05
Saturn	121,100	95.0	0.50	10^h30^m	9.54	29.5	2.5	0.06
Uranus	46,710	15.0	0.66	10^h45^m	19.2	84.0	0.8	0.05
Neptune	45,100	17.0	0.62	16^h	30.1	165.0	1.8	0.01
Pluto	<2,400	0.002	Unknown	6^d10^h	39.4	248.0	17.2	0.25

APPENDIX 3
KNOWN SATELLITES OF THE PLANETS

Planet Satellite(s)	Approx- imate Diameter (km)	Sidereal Period (days)	Mean Distance from Planet (km)	Year of Discovery	Observer
Earth					
Moon	3,480	27.3	3.9×10^5	Prehistoric	Unknown
Mars					
Phobos	24	0.3	9.3×10^3	1877	Hall
Deimos	13	1.3	2.4×10^4	1877	Hall
Jupiter					
V	150	0.5	1.8×10^5	1892	Barnard
I Io	3,640	1.8	4.2×10^5	1610	Galileo
II Europa	3,100	3.6	6.8×10^5	1610	Galileo
III Ganymede	5,270	7.2	1.1×10^6	1610	Galileo
IV Callisto	5,000	16.7	1.9×10^6	1610	Galileo
XIII	10	239	1.1×10^7	1974	Kowal
VI	120	250.6	1.1×10^7	1904	Perrine
VII	50	259.7	1.2×10^7	1905	Perrine
X	15	263.6	1.2×10^7	1938	Nicholson
XII	15	631.1	2.1×10^7	1951	Nicholson
XI	25	692.5	2.3×10^7	1938	Nicholson
VIII	50	738.9	2.4×10^7	1908	Melotte
IX	25	758.0	2.4×10^7	1914	Nicholson
XIV	Unknown	Unknown	Unknown	1975	Kowal
Saturn					
Janus	350	0.7	1.6×10^5	1966	Dolfus
Mimas	480	0.9	1.8×10^5	1789	Herschel
Enceladus	725	1.4	2.4×10^5	1789	Herschel
Tethys	965	1.9	2.9×10^5	1684	Cassini
Dione	885	2.7	3.7×10^5	1684	Cassini
Rhea	1,350	4.5	5.3×10^5	1672	Cassini
Titan	5,800	15.9	1.2×10^6	1655	Huygens
Hyperion	390	21.3	1.5×10^6	1848	Bond
Iapetus	1,210	79.3	3.5×10^6	1671	Cassini
Phoebe	290	550.5	1.3×10^7	1898	Pickering

(continued)

APPENDIX 3 (continued)

Planet Satellite(s)	Approx-imate Diameter (km)	Sidereal Period (days)	Mean Distance from Planet (km)	Year of Discovery	Observer
Uranus					
Miranda	290	1.4	1.2×10^5	1948	Kuiper
Ariel	580	2.5	1.9×10^5	1851	Lassell
Umbriel	390	4.1	2.7×10^5	1851	Lassell
Titania	970	8.7	4.3×10^5	1787	Herschel
Oberon	870	13.5	5.8×10^5	1787	Herschel
Neptune					
Triton	3,870	5.9	3.5×10^5	1846	Lassell
Nereid	290	359.9	5.6×10^6	1949	Kuiper
Pluto					
Charon	960	6.4	2.0×10^4	1978	Christy

APPENDIX 4

PHYSICAL PROPERTIES OF ORDINARY STARS

Spectral Type		Surface Temperature (°K)	Absolute Visual Magnitude	Luminosity ($L_\odot$)	Radius ($R_\odot$)	Mass ($M_\odot$)
O5	I	50,000	−10	1,000,000	—	~100
	III	50,000	−8	160,000	—	80
	V	50,000	−6	25,000	18	40
B0	I	25,000	−7	65,000	20	50
	III	25,000	−5	10,000	16	30
	V	25,000	−4	4,000	8	16
A0	I	12,000	−5	10,000	40	16
	III	12,000	−1	250	7	8
	V	12,000	0	100	3	4
F0	I	8,000	−5	10,000	65	12
	III	8,000	+1	40	8	7
	V	8,000	+3	7	1.4	2
G0	I	6,000	−5	10,000	100	10
	III	6,000	+2	16	6	3
	V	6,000	+5	1	1	1
K0	I	4,500	−5	10,000	200	12
	III	4,500	+1	40	16	4
	V	4,500	+6	0.4	0.9	0.8
M0	I	3,500	−5	10,000	500	16
	III	3,500	0	100	20	6
	V	3,500	+10	0.01	0.7	0.5

APPENDIX 5

SYSTEMS OF STARS

System	Approximate Number of Stars	Approximate Linear Diameter	Approximate Total Mass
Binary stars	2	$<1-10^4$ AU	$0.1-100 M_\odot$
Multiple stars	3–10	$10-10^5$ AU	$0.1-100 M_\odot$
Star clusters			
Associations	10–100	20–200 pc	$10-10^3 M_\odot$
Galactic clusters	50–1000	1–10 pc	$50-10^4 M_\odot$
Globular clusters	10^4-10^6	10–100 pc	$10^4-10^6 M_\odot$
Galaxies			
Irregular	10^8-10^{10}	10^3-10^4 pc	$10^8-10^{10} M_\odot$
Spiral	10^9-10^{12}	10^4-10^5 pc	$10^9-10^{12} M_\odot$
Elliptical	10^6-10^{13}	10^3-10^5 pc	$10^6-10^{13} M_\odot$
Clusters of galaxies	$10^{10}-10^{15}$	10^6-10^7 pc	$10^{10}-10^{15} M_\odot$

APPENDIX 6

THE CONSTELLATIONS

Constellation	Description	Approximate Location RA	DEC	Constellation	Description	Approximate Location RA	DEC
Andromeda	Chained Lady	1ʰ	+40°	Lacerta	Lizard	23ʰ	+45°
Antlia	Air Pump	10ʰ	−35°	Leo	Lion	10ʰ	+20°
Apus	Bird of Paradise	17ʰ	−75°	Leo Minor	Small Lion	10ʰ	+35°
Aquarius	Water Bearer	12ʰ	−05°	Lepus	Hare	6ʰ	−20°
Aquila	Eagle	20ʰ	+10°	Libra	Scales	15ʰ	−15°
Ara	Altar	17ʰ	−50°	Lupus	Wolf	15ʰ	−40°
Aries	Ram	3ʰ	+20°	Lynx	Lynx	8ʰ	+45°
Auriga	Charioteer	6ʰ	+40°	Lyra	Harp	19ʰ	+40°
Bootes	Herdsman	15ʰ	+40°	Mensa	Table Mountain	6ʰ	−75°
Caelum	Graving Tool	5ʰ	−40°	Microscopium	Microscope	21ʰ	−35°
Camelopardus	Giraffe	6ʰ	+70°	Monoceros	Unicorn	7ʰ	0°
Cancer	Crab	9ʰ	+20°	Musca	Fly	13ʰ	−70°
Canes Venatici	Hunting Dogs	13ʰ	+40°	Norma	Level	16ʰ	−50°
Canis Major	Large Dog	7ʰ	−20°	Octans	Octant	21ʰ	−85°
Canis Minor	Small Dog	8ʰ	+05°	Ophiuchus	Serpent Holder	18ʰ	0°
Capricornus	Goat	21ʰ	−20°	Orion	Hunter	6ʰ	0°
Carina	Keel of Argo	10ʰ	−60°	Pavo	Peacock	20ʰ	−65°
Cassiopeia	Queen of Ethiopia	1ʰ	+60°	Pegasus	Winged Horse	23ʰ	+20°
Centaurus	Centaur	13ʰ	−40°	Perseus	Perseus	3ʰ	+45°
Cepheus	King of Ethiopia	22ʰ	+70°	Phoenix	Phoenix	0ʰ	−45°

(continued)

APPENDIX 6 (continued)

Constellation	Description	RA	DEC
Cetus	Whale	2ʰ	−10°
Chameleon	Chameleon	11ʰ	−80°
Circinus	Compasses	15ʰ	−65°
Columba	Dove	6ʰ	−35°
Coma Berenices	Berenices Hair	13ʰ	+20°
Corona Australis	Southern Crown	19ʰ	−40°
Corona Borealis	Northern Crown	16ʰ	+30°
Corvus	Crow	12ʰ	−15°
Crater	Cup	11ʰ	−15°
Crux	Cross	12ʰ	−60°
Cygnus	Swan	21ʰ	+40°
Delphinus	Dolphin	21ʰ	+10°
Dorado	Swordfish	5ʰ	−65°
Draco	Dragon	18ʰ	+65°
Equuleus	Small Horse	21ʰ	+05°
Eridanus	River	4ʰ	−20°
Fornax	Furnace	3ʰ	−35°
Gemini	Twins	7ʰ	+20°
Grus	Crane	22ʰ	−45°
Hercules	Hercules	17ʰ	+30°
Horologium	Cloak	3ʰ	−50°
Hydra	Sea Serpent	9ʰ	−10°
Hydrus	Water Snake	2ʰ	−75°
Indus	Indian	21ʰ	−55°

Constellation	Description	RA	DEC
Pictor	Easel	5ʰ	−50°
Pisces	Fishes	1ʰ	+10°
Pisces Austrinus	Southern Fish	22ʰ	−30°
Puppis	Stern of Argo	8ʰ	−30°
Pyxis	Compass of Argo	9ʰ	−30°
Reticulum	Net	4ʰ	−60°
Sagitta	Arrow	20ʰ	+20°
Sagittarius	Archer	19ʰ	−30°
Scorpio	Scorpion	17ʰ	−35°
Sculptor	Sculptor's Tools	1ʰ	−30°
Scutum	Shield	19ʰ	−10°
Serpens	Serpent	16ʰ	+10°
Sextans	Sextant	10ʰ	0°
Taurus	Bull	4ʰ	+20°
Telescopium	Telescope	19ʰ	−50°
Triangulum	Triangle	2ʰ	+30°
Triangulum Australe	Southern Triangle	16ʰ	−65°
Tucana	Toucan	0ʰ	−65°
Ursa Major	Large Bear	11ʰ	+60°
Ursa Minor	Small Bear	16ʰ	+80°
Vela	Sail of Argo	10ʰ	−50°
Virgo	Virgin	13ʰ	0°
Volans	Flying Fish	8ʰ	−70°
Vulpecula	Fox	20ʰ	+25°

APPENDIX 7
STAR MAPS

How to use these maps: The dates on the following maps correspond to the position relative to the stars of the observer's celestial meridian (north-south line) at 8:00 PM local standard time. At 8:00 PM local time on November 1, for example, the bright star Fomalhaut will be very nearly on the observer's celestial meridian. To find the location of the celestial meridian for times other than 8:00 PM, move the meridian one hour of right ascension per hour of time—west for times earlier than 8:00 PM, east for times later than 8:00 PM. Thus the meridian is located at 22^h at 7:00 PM on November 1; at 0^h (24^h) at 9:00 PM, and so on. Stars having a declination equal numerically to the observer's latitude will pass directly overhead.

NORTH CIRCUMPOLAR STARS

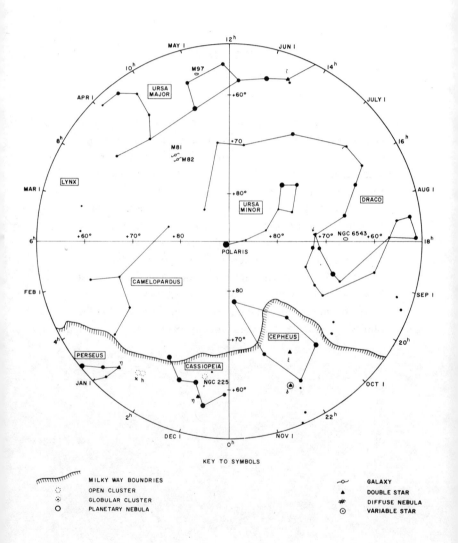

KEY TO SYMBOLS

∿∿∿	MILKY WAY BOUNDRIES	∿⊸	GALAXY
∵	OPEN CLUSTER	▲	DOUBLE STAR
⊛	GLOBULAR CLUSTER	✳	DIFFUSE NEBULA
O	PLANETARY NEBULA	⊙	VARIABLE STAR

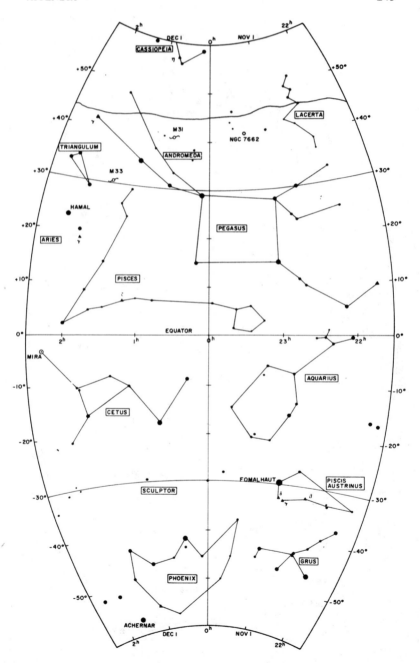

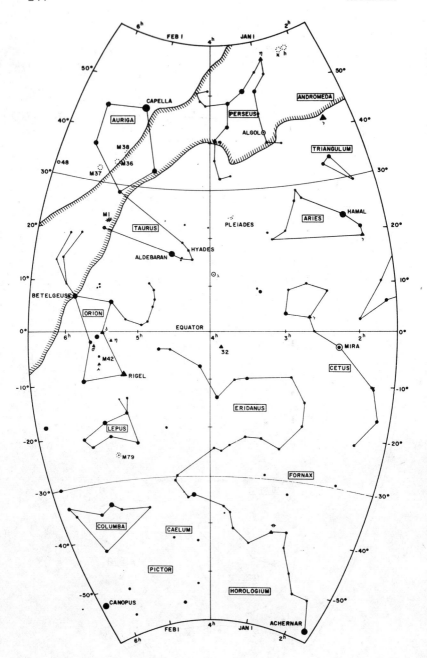

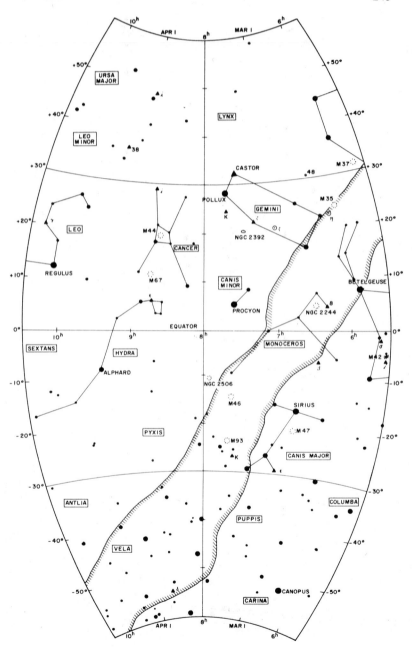

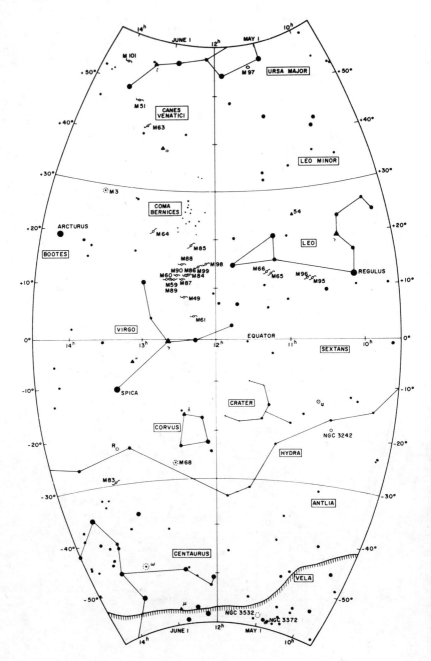

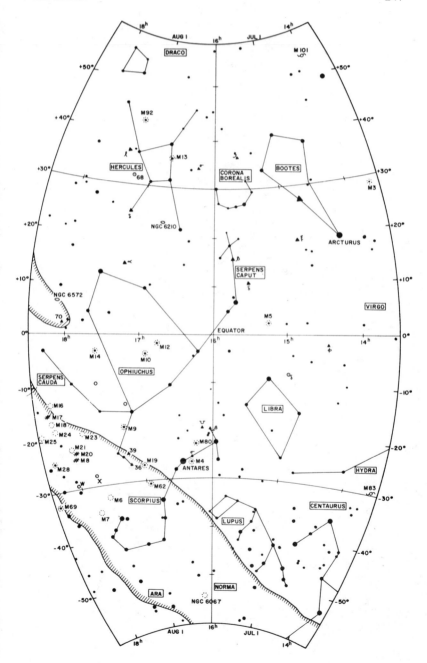

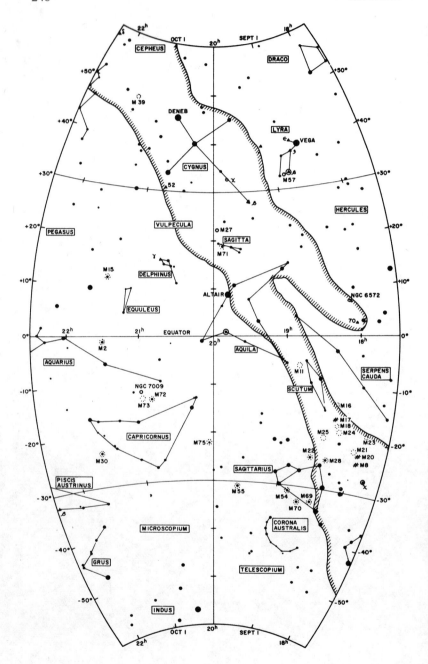

SOUTH CIRCUMPOLAR STARS

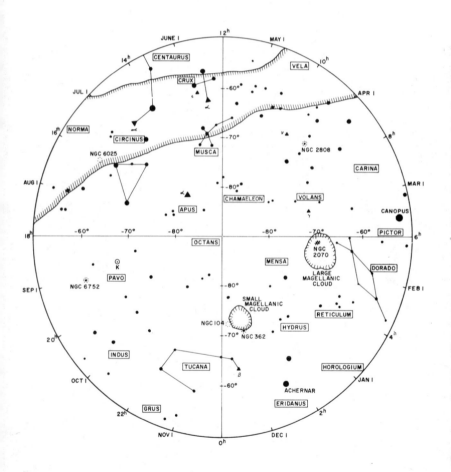

Glossary

A-type star One of a class of stars with a temperature between 8,000° and 12,000°K and spectrum characterized by strong hydrogen lines.

aberration A defect in the image produced by a poorly figured telescope lens or mirror.

aberration of starlight The apparent displacement of a star's observed position due to the motion of the earth around the sun.

absolute magnitude The magnitude that a given object would appear to have if viewed from a distance of 10 parsecs.

absolute temperature A scale of temperature in which the zero point is the lowest possible temperature, $-273°C$.

absolute zero The temperature at which all molecular motion ceases; 0°K or $-273°C$.

absorption line A narrow-wavelength region of greatly reduced intensity in a continuous spectrum.

absorption spectrum A series of dark-absorption lines superimposed on a continuous spectrum.

acceleration A change in the magnitude and/or direction of a velocity.

accretion The process by which small particles collide and stick together to form larger masses.

active sun   The sun at a time when it exhibits relatively large numbers of sunspots, prominences, flares, and other forms of atmospheric activity.

aerolite   A stony meteorite.

airglow   A constant background of light caused by fluoresence in the upper layers of the earth's atmosphere.

albedo   A measure of the ability of a planet, satellite, or meteoroid to reflect sunlight.

almanac   A tabulation of astronomical events.

alpha particle   An atomic particle consisting of two protons and two neutrons; the nucleus of a helium atom.

altitude   The angular distance, measured along a vertical circle, between an object and the horizon as seen by the observer.

amplitude   The range of variability, such as light from a star.

Angstrom (Å)   A very small unit of length used to measure the wavelengths of visible light rays; one Angstrom is equal to 10^{-8} cm.

angular diameter   The angle subtended by an object's diameter as viewed from a given distance to the object.

angular momentum   The momentum of an object or particle moving about a point or axis.

annular eclipse   A solar eclipse in which the moon's angular diameter is less than the sun's, resulting in a donut-like appearance for the sun during the eclipse's total phase.

anomalistic month   The interval of time between two successive passages of the moon through perigee.

anomalistic year   The interval of time between two successive passages of the earth through its perihelion point.

antapex   The point on the celestial sphere that is exactly opposite to the direction of the sun's motion.

apex   The point on the celestial sphere toward which the sun is moving.

aperture   The "hole" through which light enters a telescope or camera.

aphelion   For an object orbiting the sun, the point on the orbit farthest from the sun.

apogee   For an object orbiting the earth, the point on the orbit farthest from the earth.

apparent magnitude   The brightness of an object as seen from the earth.

apparent solar day   The interval of time between two successive passages of the sun's center across the celestial meridian.

apsidal motion   The change of the space orientation of the major axis, the line joining the near and far points of an object's orbit.

arcminute   An angle equal to $^1/_{60}$ of a degree.

arcsecond   An angle equal to $^1/_{60}$ of an arc minute.

ascending node   Point at which an object's orbit crosses a given reference plane in a specified "upward" direction.

association (stellar)   A loose collection of stars that have a common origin.

asteroid   One of several thousand relatively large bodies that usually have all or a part of their orbital paths between Mars and Jupiter.

asteroid belt   Region of the solar system between the orbits of Mars and Jupiter in which almost all of the asteroids can be found.

astigmatism   An optical defect in which rays striking a lens or mirror in different planes are not focused at the same spot.

astrology   A branch of the occult in which configurations of the sun, moon, and planets are believed to influence human affairs.

astrometry   The branch of astronomy dealing with the precise determination of celestial positions.

astronomical unit (AU)   The mean distance between the earth and the sun; one astronomical unit equals 1.495985×10^8 km.

astronomy   The branch of science dealing with the nature of the universe beyond the atmosphere of the earth.

astrophysics   The branch of astronomy dealing with the physics of celestial objects.

atmosphere   A gaseous envelope surrounding a planet or star.

atom   The smallest division of an element that will retain the properties of that element.

atomic mass unit (amu)   A unit of mass roughly equal to the mass of a hydrogen atom or 1.67×10^{-24} g.

atomic number   The number of protons or electrons in one atom of an element in its normal state.

atomic weight   The mass of an atom expressed in atomic mass units.

aurora   A glow in the upper atmosphere of the earth emitted by atoms and ions.

aurora australis   An aurora in the Southern Hemisphere; the "southern lights."

aurora borealis   An aurora in the Northern Hemisphere; the "northern lights."

autumnal equinox   That point on the celestial equator where the sun crosses it moving from north to south; the time at which the sun crosses this point.

azimuth   The eastward angle between the vertical circle containing the north point and the vertical circle containing an object.

B-type star   One of a class of stars with a temperature between 12,000° and 25,000°K and a spectrum characterized by lines of hydrogen and neutral helium.

Bailey's beads   Small "beads" of sunlight visible around the mountainous lunar disk just before and just after the total phase of a solar eclipse.

Balmer lines   Absorption or emission lines in the visible region of the electromagnetic spectrum that arise, respectively, from transitions up from or down to the second electronic energy level of the hydrogen atom.

barium star   A G, K, or earlier M star having overabundances of the heavier metals such as barium and zirconium.

barred spiral   A spiral galaxy in which the spiral arms begin at the ends of a "bar" passing through the nucleus.

barycenter   The center of mass for a two-body system.

Be star   A B-type star with emission lines in its spectrum.

beta particle   A highly energetic electron emitted by a radioactive substance.

"Big Bang" theory   A cosmological model in which the observed expansion of the universe is set in motion by a huge primeval explosion.

binary star   A double star system whose components orbit each other, each one trapped by the other's gravity.

black dwarf   One of the possible endpoints of a star's evolution in which a star no longer generates energy and has assumed a stable configuration.

black hole   One of the possible endpoints of a star's evolution in which a star's mass becomes so compacted through gravitational collapse that neither energy nor matter can escape from it.

blackbody   An idealized object that is capable of absorbing and re-emitting all radiant energy that falls on it.

blink microscope or comparator   An instrument that allows the user to view two photographic plates alternately in rapid succession to detect any motion or light variability.

Bode's law   A numerical progression that reproduces the approximate spacing of the planets in order from the sun.

bolide   A fireball that explodes with an audible sound.

bolometric correction   The difference between the visual and bolometric magnitudes of a given object.

bolometric magnitude   The total energy output of an object at all wavelengths expressed as a magnitude.

bright-line spectrum   An emission spectrum consisting of a series of bright lines superimposed on a dark continuum.

brightness   The energy per unit area received from a given celestial object at a given distance.

C star   See carbon star.

carbon cycle   A series of nuclear reactions in which carbon nuclei act as a catalyst in the fusion of hydrogen into helium.

carbon star   A star whose spectrum is dominated by molecular bands of carbon compounds such as CN, CH, and C_2.

cardinal points   The four principal directions on a compass: north, south, east, and west.

Cassegrain telescope   A reflecting telescope in which the light is brought to a focus through a small hole in the center of the objective mirror by means of a convex secondary mirror.

Cassini's division   The most notable gap in the ring system surrounding the planet Saturn.

celestial equator   The great circle on the celestial sphere 90° from the celestial poles; the intersection of the celestial sphere and the plane of the earth's equator.

celestial horizon   The great circle on the celestial sphere halfway between the observer's zenith and nadir.

celestial mechanics   The branch of astronomy dealing with the motions and gravitational influences of celestial objects.

celestial meridian   A great circle on the celestial sphere that passes through the celestial poles and the observer's zenith.

celestial poles   The points of intersection of the earth's polar axis with the celestial sphere.

celestial sphere   An imaginary sphere of large radius centered on the earth and onto which the positions of celestial objects are projected for the purpose of measurement.

center of gravity or center of mass   The point in a single body or within a system of bodies that behaves as though the entire mass of the configuration were concentrated at that point.

centrifugal force   An imaginary force that is assumed to push a mass radically outward from the center of a given circular motion.

centripetal force   A force directed radially inward from the center of a given circular motion; the force that causes the object to move in a curved path.

cepheid variable   A class of F and G supergiant stars whose brightness and temperature vary with a regular periodicity because of internal pulsations.

Chandrasekhar limit   The largest mass (1.4 solar masses) that a white dwarf star can have without undergoing gravitational collapse.

chromatic aberration   A lens defect in which different colors of light have different focal points.

chromosphere   The layer of solar atmosphere between the photosphere and the corona.

chronograph   An instrument for recording the times of events.

chronometer   A highly accurate clock.

circular velocity   The velocity needed for a body to maintain a circular orbit about a second body.

circumpolar stars   Stars which are always above the horizon as seen from a given latitude.

cluster   Any physical grouping of stars or galaxies.

cluster variable   A class of pulsating variable stars having periods of less than one day; an RR Lyrae star.

collimator   A lens system that converts a diverging light beam into a parallel beam.

color excess   The increase in the color index of a celestial object due to light scattering by interstellar material.

color index   The difference between the blue magnitude and the yellow (visual) magnitude of a celestial object; the difference between any two magnitudes of a single object measured at different wavelengths.

color-magnitude diagram   Plot of apparent magnitude versus color index for a star cluster.

coma (comet)   The diffuse gaseous material that surrounds the nucleus of a comet head.

coma (optics)   The inability of a lens or mirror to focus off-axis rays at the focal point.

comet   A conglomeration of dust, ice, and frozen gases that orbits the sun.

comparison spectrum   An emission spectrum from a known substance placed beside the spectrum of a celestial object.

conic section   The curve of intersection between a right circular cone and a plane; can be a circle, ellipse, parabola, or hyperbola.

conjunction The closest apparent approach of one celestial object to another as seen from the earth.

constellation One of 88 defined zones in the sky named for various people, animals, or objects.

contacts The exact instants of certain stages of an eclipse.

continental drift The theory that the continents on the earth are moving away from each other at a very slow rate.

continuous spectrum The distribution by wavelength of the energy emitted by an incandescent object.

convection The transport of energy by moving currents of material.

core The central regions of a celestial object.

Coriolis effect The apparent deflection observed from the earth's surface of a projectile moving above the rotating earth.

corona The outer atmospheric layers of the sun.

coronagraph An instrument for observing the sun's outer atmospheric layers at times other than total eclipse.

corpuscular radiation Charged atomic particles continuously emitted by the sun.

cosmic rays High-energy atomic particles (mostly protons) that strike the earth's atmosphere.

cosmogony The branch of astronomy that deals with the origin and evolution of the solar system or of the universe.

cosmology The branch of astronomy that deals with the large-scale properties and structure of the universe.

coudé focus A focus point in a reflecting telescope that does not change its position relative to the earth's surface.

crater A circular depression on the surface of an object caused by either internal volcanic activity or meteoric impacts.

crescent Any phase of the moon or planets in which the observer sees less then 50 percent of the illuminated portions of the disk.

crust The outermost solid layers of a planet, satellite, or meteoroid.

dark nebula A cloud of interstellar dust that obscures the light from the stars behind it.

declination The angular distance, measured along an hour circle, between an object and the celestial equator; an "astronomical latitude."

deferent In the geocentric planetary system, a circular orbit centered on the earth along which an object or the center of the object's epicycle moves.

degenerate matter   Matter in which the electrons are compacted into their lowest possible energy levels.

density   The amount of mass contained in a unit volume of an object or a substance.

descending node   The point at which an orbit crosses a given reference plane in a specified "downward" direction.

deuterium   An isotope of hydrogen having a proton and a neutron in its nucleus; "heavy" hydrogen.

deuteron   The nucleus of a deuterium atom.

differential rotation   Rotational motion in which different parts of the object or system move at different rates from one another.

diffraction   A process by which light waves are spread out as they pass an opaque edge.

diffraction grating   An array of closely spaced lines that reflects or transmits light waves in different directions according to their wavelengths.

diffraction pattern   A series of alternating bright and dark areas (fringes) produced by the diffraction and subsequent "self-interference" of light waves.

diffuse nebula   An irregular luminous cloud of interstellar material.

disk   The apparent surface of a celestial object.

disk of Galaxy   The flat, circular regions of the Milky Way Galaxy.

dispersion   The process by which light waves are sorted out according to wavelength.

distance modulus   The distance to an object expressed as a difference between its apparent and absolute magnitude.

diurnal   Occurring on a daily basis.

diurnal circle   The apparent path of a star on the celestial sphere due to the earth's daily rotation.

diurnal parallax   The apparent change in direction of an object due to "displacement" of the observer by the rotation of the earth.

Doppler effect (shift)   The change in the observed wavelength of radiation due to relative motion between the wave source and the observer.

draconitic month   The interval of time between successive passages of the moon through its ascending (or descending) node.

dwarf   Term applied to a star on the main sequence, especially at the low luminosity end.

dynamic parallax   A parallax for a binary star system derived from Kepler's harmonic law, the mass-luminosity law, and the period of mutual revolution.

dynamo theory   Theory that the magnetic field of a planet arises because of its rotation about a liquid metallic core of significant size.

dyne   The unit of force in physics that is required to accelerate 1 gram of mass 1 centimeter per second per second.

earthshine (or earthlight)   The sunlight reflected from the earth that dimly illuminates the dark portions of the moon near times of new moon.

east point   The point on the horizon 90° to the right of the north point.

eccentric   In the geocentric planetary system, the center of a circular orbit of the sun that is ''offset'' from the earth.

eccentricity   A measure of the shape of an ellipse; the ratio of the distance between the foci to the major axis.

eclipse   The cutting off of the light of a celestial object by another body passing in front of it.

eclipsing binary   A binary star with light variations produced by one member eclipsing the other.

ecliptic   The apparent annual path of the sun among the stars; the circle of intersection between the earth's orbital plane and the celestial sphere.

effective temperature   The temperature to which a blackbody of equal size must be raised in order to produce the observed luminosity of the given object.

electromagnetic radiation   Radiant energy produced by oscillating electric or magnetic charges or fields.

electromagnetic spectrum   The sum total of all known wavelengths of electromagnetic radiation including gamma rays, X rays, ultraviolet rays, visible light, infrared, and radio waves.

electron   A subatomic particle with a negative electronic charge that occupies the outer regions of an atom.

element (chemical)   Any substance that cannot be broken down further by normal chemical reactions.

element (orbital)   One of a set of parameters used to describe the orbit of an object in space.

ellipse   A conic section; the curve of intersection of a circular cone and a plane cutting through it.

elliptical galaxy   A galaxy having an elliptical shape but no trace of spiral structure.

elongation   The apparent angular separation between an object and the sun as seen from the earth.

emission line   A bright line in a spectrum caused by electrons making transitions to lower energy levels, thereby releasing energy in the form of photons.

emission nebula   A nebula that radiates visible light by fluorescing the ultraviolet light from stars within or near the nebula.

emission spectrum   A spectrum consisting of emission lines.

encounter   A chance close approach of two nonorbiting objects that produces gravitational effects on both objects.

energy   The ability to do work.

energy levels of atoms   The possible energies of electronic configurations of atoms above the least energetic configuration or ground state.

ephemeris   A table listing the positions of a celestial object at specified times.

epicycle   In the geocentric planetary system, a small circular orbit whose center moves along the circumference of a deferent.

epoch   A date specified as a time reference for astronomical observations.

equant   In the geocentric planetary system, a noncentral point associated with a circular orbit about which an object moves at constant angular velocity.

equation of state   An equation that expresses the relationship among the pressure, temperature, and density of a substance.

equation of time   The difference between apparent and mean solar time.

equator   The great circle on the earth's surface located 90° from the celestial poles.

equatorial mounting   A telescope mounting whose two axes of rotation are aligned perpendicular to and parallel with the earth's axis of rotation.

equatorial system of coordinates   A celestial coordinate system using the celestial equator as its primary reference plane.

equinox   One of two points of intersection on the celestial sphere between the ecliptic and the celestial equator.

equivalent width   A measure of the strength of a spectral line.

erg   The amount of work done by a force of 1 dyne moving through a distance of 1 centimeter; the unit of energy in the centimeter-gram-second system of metric units.

eruptive variable   A variable star whose light changes are sudden and erratic.

escape velocity   The minimum velocity needed for a given object to escape the gravitational field of a second object.

eyepiece A small lens used to examine the images produced by the primary lens or mirror of a telescope.

excitation The increasing of the energy of an atom's electronic configuration through collisions or by absorption of radiant energy.

exosphere The outermost layer of the earth's atmosphere.

extinction The dimming of light as it passes through material that will scatter or absorb it.

extragalactic Outside of the boundaries of the Milky Way Galaxy.

f number See focal ratio.

F-type star One of a class of stars with a temperature between 6000° and 8000°K and a spectrum characterized by lines of singly ionized metals.

faculae Bright regions on the sun near the solar limb.

filament The dark silhouette of a prominence on the solar disk.

filar micrometer An auxiliary instrument used with a telescope to measure precisely relative orientations, angular sizes, and separations of celestial objects.

fireball An unusually bright meteor.

first quarter The lunar phase that occurs when the moon is 90° east of the sun as seen from the earth.

fission The breaking up or "smashing" of heavier nuclei into lighter ones with an attendant release of energy.

flare A sudden, temporary increase in the brightness of a localized region on the sun accompanied by a burst of high-energy particles.

flare star A variable star, usually a red dwarf, that suddenly and unpredictably increases its brightness for brief periods of time.

flash spectrum The spectrum of the solar chromosphere that manifests itself for a brief instant just as totality begins.

flocculi Bright regions in the magnetic fields surrounding sunspots that are visible on spectroheliograms of the sun; plages.

fluorescence The absorbtion of light or energy at short wavelengths and its subsequent reemission at other, longer wavelengths.

focal length The distance between a lens or mirror and its focus.

focal ratio The ratio of the size of the aperture of a lens or mirror to its focal length.

focus The point in an optical system where the image is formed.

forbidden lines Spectral lines not usually observed in a laboratory because they result from electronic transitions that are highly unlikely.

force   Any agent in nature that will produce or prevent the motion of an object.

Fraunhofer line   An absorption line in the spectrum of the sun or a star.

Fraunhofer spectrum   An array of absorption lines in the spectrum of the sun or a star.

free-free transition   An interaction between an ion or atom and a passing electron in which energy of motion is exchanged and radiant energy is either absorbed or emitted without the electron being captured.

frequency   The number of oscillations in a wave motion passing a given point per unit time.

fringes   The set of light and dark regions caused by the diffraction and subsequent interference of light waves as they pass an opaque edge.

full moon   The lunar phase that occurs when the moon is exactly opposite the sun as seen from the earth.

fusion   The nuclear "melting" of lighter elements into heavier ones with an attendant conversion of mass into energy.

G-type star   One of a class of stars with a temperature between 4500° and 6000°K and a spectrum characterized by lines of neutral and singly ionized metals; a solar-type star.

galactic cluster   A loose collection of stars located in the disk of the Milky Way.

galactic equator   The circle of intersection of the plane of the Milky Way disk with the celestial sphere.

galactic latitude   The shortest angular distance between the galactic equator and a given object.

galactic longitude   The angular distance measured eastward from the galactic center to the point of intersection between the galatic equator and the galactic meridian of the object.

galactic meridian   A great semicircle having the galactic poles as its endpoints.

galactic poles   The two points on the celestial sphere 90° from the galactic equator.

galactic rotation   The rotational motion of the Milky Way Galaxy about its center.

Galaxy   The Milky Way star system.

galaxy   One of a multitude of remote star systems containing billions of stars.

Galilean satellites The four largest satellites of Jupiter, which were discovered by Galileo.

gamma rays That part of the electromagnetic spectrum having wavelengths shorter than 1 Å; the most energetic region of the electromagnetic spectrum.

gas giant One of the low-density planets: Jupiter, Saturn, Uranus, or Neptune.

gauss A unit of magnetic field intensity in the centimeter-gram-second system of measurement.

gegenshein A faint, diffuse glow in the sky opposite the sun; the counterglow.

geocentric Having the earth at the center.

geodesy The branch of earth science that deals with the measurement of the earth's size and shape.

geomagnetic Of or referring to the earth's magnetic field.

giant branch A sequence on the H-R diagram occupied by stars that have ended the main-sequence phases of their evolution.

giant star A star having a large luminosity or radius.

gibbous Any phase of the moon or planets in which the observer views between 50 and 100 percent of an object's illuminated surface.

globular cluster One of the large, tightly packed spherical systems of stars that occupy a roughly spherical distribution relative to the center of a galaxy.

globule A small, compact dark nebula believed to be a star in the process of formation.

Gondwanaland A hypothesized primordial land mass in the Southern Hemisphere that broke up to form some of our modern continents and islands.

granules The pattern of convective cells in the solar photosphere responsible for the sun's mottled or granular appearance.

gravitation The fundamental property of a mass by which it exerts a force of attraction on any other mass.

gravitational collapse The collapse of a mass configuration by self-gravity.

gravitational redshift The longward shift to wavelength experienced by a photon as it attempts to leave the surface of a given mass.

great circle Any circle on the surface of a sphere whose center coincides with that of the sphere.

greatest elongation   The maximum angular separation between the sun and either Mercury or Venus as seen from the earth.

greenhouse effect   The trapping of a planet's radiant energy by its atmosphere.

Greenwich meridian   The meridian that passes through Greenwich, England, and by international agreement is the zero point for longitude measurement; the prime meridian.

Greenwich time   The standard time at the Greenwich meridian.

Gregorian calendar   The calendar introduced in 1582 by Pope Gregory XIII and which is the most commonly used calendar today.

ground state   The lowest possible energy for an atom's electronic configuration.

H I region   A region of neutral hydrogen in space.

H II region   A region of ionized hydrogen in space.

H line   The strong Fraunhofer line of ionized calcium at 3968 Å.

half-life   The time required for one-half of the atoms in a radioactive substance to decay.

halo (atmospheric)   A ring of light around the sun or moon produced from the refraction of light by high-altitude ice crystals.

halo (galactic)   The stars, star clusters, and other material spherically distributed about the nucleus of a galaxy.

harmonic law   Kepler's law of planetary motion, which states that the ratio of the cube of the mean distance to the square of the sidereal period is constant for every planet.

harvest moon   The full moon that occurs nearest the time of the autumnal equinox.

Hayashi tracks   The theoretical evolutionary path of the H-R diagram of a convective star in its formative stages.

heavy element   An element whose atomic number is greater than that of the element helium (2).

heliocentric   Having the sun at the center.

helium flash   An explosive ignition of the helium in the core of a red giant that starts the fusion of helium into carbon.

Herbig-Haro object   A pre-main sequence stage of a star's evolution, thought to occur after the globule stage and before the star becomes a T Tauri star.

Hertzsprung gap   A triangular region in the upper central part of the H-R diagram where few stable stars are found.

Hertzsprung-Russell (H-R) diagram   A plot of the absolute magnitudes of a group of stars against their temperature, spectral class, or color index.

high-velocity star　A star having a large space motion relative to the sun as a result of its not sharing the galactic orbital motion of the sun.

horizon　A great circle on the celestial sphere 90° from the observer's zenith.

horizon system　A system of celestial coordinates having the horizon as its principal reference plane.

horizontal branch　A sequence of stars on the H-R diagram the members of which have absolute magnitudes roughly equal to zero and which are believed to have passed through the red-giant stage of evolution.

horizontal parallax　A parallax obtained by using the earth's equatorial radius as a baseline.

hour angle　The angle between the hour circle of the object and the celestial meridian.

hour circle　A great circle on the celestial sphere that passes through the celestial poles.

H-R diagram　See Hertzsprung-Russell diagram.

Hubble constant　The constant of proportionality in the Hubble law; its value is of the order of 50 km/sec/megaparsec.

Hubble's law　The statement that the observed radial velocity of a receding galaxy is proportional to its distance.

hydrostatic equilibrium　A condition in an object in which inward gravitational forces are exactly balanced by the forces pushing outward.

hyperbola　The geometric figure generated by slicing a right circular cone parallel to its axis with a plane; the shape of an encounter orbit.

igneous rock　A rock formed from molten material.

image　The rendition of an object by an optical system.

image tube　A device that enhances the brightness of an image through the use of photoelectric processes.

inclination　The angle between the plane of an object's orbit and a specified reference plane such as the plane of the celestial equator or of the ecliptic.

Index Catalogue (IC)　A supplement to the New General Catalogue (NGC) of diffuse objects.

index of refraction A measure of the ability of a substance to refract light rays; the ratio of the speed of light in a vacuum to its speed in a given substance.

inertia The property of matter by which an object resists any attempt to change its velocity.

inferior conjunction A conjunction of an inferior planet and the sun when the planet is between the earth and sun.

inferior planet A planet whose orbital radius is less than that of the earth; the planets Mercury or Venus.

infrared radiation That part of the electromagnetic spectrum from 7000 Å to about 1 mm; wavelengths longer than the ones we perceive as red "heat" waves.

insolation The amount of solar radiation that falls on a unit area of the earth's surface per unit time.

interferometer An optical device that measures small angular distances using the principle of interference of light waves.

International Date Line A line that roughly coincides with the 180° meridian on the earth's surface across which the date changes by a day.

international magnitude system An outdated system of magnitudes based on brightnesses recorded on blue- and yellow-sensitive photographic plates.

interplanetary medium The distribution of gas and dust in interplanetary space.

interstellar dust Microscopic solid grains in interstellar space.

interstellar gas The diffuse gas in interstellar space.

interstellar lines Absorption lines produced by interstellar gas superimposed on stellar spectra.

interstellar medium The distribution of gas and dust in interstellar space.

interstellar reddening The reddening of starlight caused by scattering from the dust particles in the interstellar medium.

ion An atom that has a nonzero net electronic charge.

ionization The process by which a neutral atom gains or loses electrons.

ionosphere The upper layer of the earth's atmosphere characterized by a high percentage of ionized oxygen and nitrogen.

irons (siderites) A class of meteorite composed of 90 percent iron, 9 percent nickel, and 1 percent other materials.

irregular galaxy A galaxy that lacks symmetry.

irregular variable A variable star whose light variations are not periodic.

island universe An outdated term for galaxy.

isotopes Atoms of the same element that have the same atomic numbers but different atomic weights.

isotropy The property of space such that one direction is the same as any other.

Jovian planet Any of the gas giants: Jupiter, Saturn, Uranus, or Neptune.

Julian calendar A calendar invented by Julius Caesar in 46 B.C.

Juno Third asteroid to be discovered.

K-line A rather strong Fraunhofer line at 3933 Å due to ionized calcium.

K-type star One of a class of stars with a temperature between 3500° and 4500°K and a spectrum characterized by weak molecular bands and absorption lines from neutral metals.

Kepler's laws The three basic laws governing the motions of the planets put forth by Kepler early in the seventeenth century.

kiloparsec One thousand parsecs.

kinetic energy The energy associated with an object's motion.

kinetic theory A description of fluids that seeks to explain fluid properties in terms of molecular motions.

Kirchoff's laws The three laws that explain the conditions under which continuous, bright-line (emission), and dark-line (absorption) spectra are formed.

Kirkwood's gaps Gaps in the orbital spacing of the asteroids that arise from gravitational perturbations of the planets, especially Jupiter.

Lagrangian points Five points in the plane of two orbiting bodies at which the net forces from the two bodies are zero.

latitude The shortest angular distance between a location on the earth's surface and the earth's equator as seen from the earth's center.

law A statement describing the behavior of a phenomenon in nature.

law of equal areas Kepler's second law of planetary motion: the line joining the sun and a planet (radius vector) sweeps out a constant orbital area per unit time.

law of the red shifts Hubble's law.

leap year A 366-day calendar year employed every fourth year divisible by four in order to keep the civil calendar in phase with the solar year.

libration Small periodic changes in the relative orientation between an observer on the earth and the moon that allows one to see more than a hemisphere of the lunar surface.

light That part of the electromagnetic spectrum (roughly 4000 Å to 7000 Å) that can be seen with the human eye.

light curve A plot of an object's apparent brightness versus time.

light-year The distance light travels in 1 year, or 9.5×10^{12} km.

limb The edge of a celestial object's apparent disk.

limb darkening The phenomenon whereby the center of the disk of a celestial object is brighter than its limb regions.

limiting magnitude The faintest magnitude that can be seen with a given instrument under given observing conditions.

line broadening An effect in which spectral lines are spread out over a range of wavelengths owing to a variety of physical processes.

line of apsides The line joining the near and far points of an orbit; the major axis of an orbit.

line of nodes The line joining the nodal points of an orbit.

line profile A detailed plot of intensity versus wavelength for a spectral line.

linear diameter The actual diameter of an object in units of distance or length.

local apparent time The hour angle of the sun.

Local Group A small cluster of some twenty galaxies to which the Milky Way Galaxy belongs.

local standard of rest A coordinate system that shares the average motion of the sun and nearby stars about the galactic center.

longitude The angular distance between the meridian of a given location and the Greenwich meridian as seen from the earth's center.

long-period variable A variable star whose brightness changes occur with periods longer than 70 days.

low-velocity star A star having a very small space motion relative to the sun and hence believed to be a part of the general galactic rotation.

luminosity The total amount of energy given off by an object per unit time.

luminosity class The classification assigned to a star given spectral type on the basis of its luminosity.

luminosity function The relative numbers of stars having various absolute magnitudes per unit volume of space.

lunar Of or pertaining to the moon.

Lyman series The series of spectral lines of hydrogen that arise from electronic transitions into and out of the lowest energy level of the hydrogen atom.

M-type star One of a class of stars with a temperature lower than 3500°K and a spectrum characterized by molecular bands of titanium oxide.

Magellanic Clouds A pair of irregular galaxies which are satellite galaxies of the Milky Way Galaxy.

magnetic field A region of space within which magnetic forces can be detected.

magnetic poles One of two points on a body at which the magnetic forces are strongest.

magnetosphere A region around a planet in which its magnetic field strongly affects the motions of incident charged particles.

magnification The apparent size of an object seen through an optical system compared to its size when viewed with the unaided eye.

magnitude A scale of measuring brightness in which each magnitude jump represents a 2.5-factor increase or decrease in brightness.

main sequence A sequence of stars on the H-R diagram to which the vast majority of stars belong.

major axis The longest line that can be drawn between two points on a closed orbit.

major planet Any one of the nine planets in the solar system.

mantle The layer of material between a planet's core and its crust.

mare (plural maria) A dark sea-like planar area on the moon's surface.

mascons Localized concentrations of mass, especially just below the lunar surface.

mass A measure of the amount of matter in an object defined by its inertial or gravitational properties.

mass defect The difference between the atomic number and the mass of an atomic nucleus.

mass function In a single-line spectroscopic binary, the observed ratio of the cube of the product of the second object's mass and the sine of its angle of orbital inclination to the square of the sum of the two masses.

mass-luminosity relation   The correlation between stellar mass and stellar luminosity, especially for stars on the main sequence.

mass-radius relation   The correlation between stellar mass and stellar radius, especially for white dwarf stars or stars on the main sequence.

mean solar day   The interval between successive passages of the sun across the celestial meridian, assuming that the apparent eastward motion of the sun is constant throughout the year.

mean solar time   The hour angle of the sun plus 12 hours.

megaparsec   One million parsecs.

meridian   A great circle on the earth's surface that passes through the north and south poles.

meson   A subatomic particle having a mass between that of a proton and an electron.

mesophere   The layer of the ionosphere that lies immediately above the stratosphere.

Messier catalogue   A catalogue of about 100 diffuse objects compiled by Charles Mesier in the eighteenth century.

metal   In astronomy a loose term denoting any element that is not hydrogen or helium.

metastable level   An energy level in an atom from which there is a very low probability that an electron will make a photon-generating transition.

meteor   The luminous event in which a meteoroid is burned up in the earth's atmosphere; a "falling" or "shooting" star.

meteor shower   A celestial event in which an uncommon number of meteors appear to radiate from a specific point in the sky.

meteor stream   A group of meteoroids uniformly distributed along a highly elliptical orbit, usually that of an old comet.

meteor swarm   A group of meteoroids clustered about a single point moving along a highly elliptical orbit, usually that of an old comet.

meteorite   The portion of a meteoroid that survives passage through the earth's atmosphere and strikes the earth.

meteoroid   Any solid interplanetary particle that is not a major planet or satellite of a major planet.

micrometeorite   An extremely small meteoroid that because of its small size can filter through the earth's atmosphere to the ground without burning up.

microphotometer   A device for accurately measuring the photographic density of the image(s) recorded on a photograph or spectrogram.

microwaves That part of the radio region of the electromagnetic spectrum having wavelengths roughly between 1 mm and 30 cm.

Milky Way Galaxy The vast spiral-shaped assemblage of stars to which the sun belongs and which we see as a dim band of light encircling the celestial sphere.

minor axis The shortest line that can be drawn from one point on an orbit through the geometric center of the orbit to a second point on the orbit.

minor planet One of several tens of thousands of objects in the solar system having diameters larger than 1 km but that are neither one of the nine principal planets nor one of their satellites.

Mira variables A red giant variable M-type star whose light variation occurs over periods of more than 70 days; the star Mira (o Ceti) is the prototype.

model atmosphere A theoretical calculation of the physical conditions in any layer of the atmosphere of an object, especially a star.

model interior A theoretical calculation of the physical conditions in any layer of the interior of an object, especially a star.

molecular band A set of emission or absorption lines arising from a molecule that are so closely spaced that they blend into a single broad spectral feature.

molecule The smallest division of a compound that will retain all of the chemical properties of that compound.

momentum The product of an object's mass and velocity.

monochromatic Consisting of a single color or wavelength.

moon The satellite of the earth; any lesser body in orbit about a planet.

N galaxy A galaxy having an extremely bright nucleus.

nadir The point on the celestial sphere directly opposite the observer's zenith.

nautical mile The actual distance between two locations on the same meridian separated by one arcminute in latitude.

neap tide The lowest tides in a given month; neap tides occur around the time of quarter moon.

nebula A cloud of interstellar material.

neutrino A subatomic particle that has no mass or charge.

neutron A subatomic particle that has about the same mass as a proton but has no electronic charge.

neutron star One of the final stages of stellar evolution in which protons and electrons have all been gravitationally crushed into neutrons.

New General Catalogue A catalogue of diffuse objects that was compiled by Dreyer in 1888 and succeeded the Messier catalogue.

new moon The lunar phase that occurs at a conjunction of the sun and moon.

Newtonian telescope A reflecting telescope in which the light rays are brought to a focus point at the sides of the main tube through the use of a small, flat secondary mirror.

Newton's laws The three basic statements in Newton's description of motion.

node One of two points of intersection between an orbit and some specified reference plane.

nodical month The time for one revolution of the moon relative to one of its nodes.

north point The point of intersection between the horizon and the celestial meridian that is closest to the north celestial pole.

north polar sequence A group of stars near the north celestial pole used to define the zero point of the magnitude scale.

nova A star that undergoes a large, sudden increase in its total energy output, probably from an internal nuclear explosion.

nuclear fission The process in which atomic nuclei are broken apart into smaller nuclides with an attendant conversion of mass into energy.

nuclear fusion The process in which lighter atomic nuclei are "melted" into heavier nuclides with an attendant conversion of mass into energy.

nucleosynthesis The building up of atomic nuclei by nuclear reactions.

nuclide An atomic nucleus.

nucleus The central region of an atom, a comet, or a galaxy.

nutation A nodding perturbation on the precessional motion of earth's axis of rotation due to the gravitational pull of the moon on the earth's equatorial bulge.

O-type star One of a class of stars having a temperature higher than 35,000°K and a spectrum characterized by lines of ionized helium and highly ionized metals.

objective The main light-gathering lens or mirror of a telescope.

objective prism A lens placed in front of a telescope objective that converts each image in the field of view into its spectrum.

oblate spheroid A solid figure formed by rotating an ellipse about its minor axis.

oblateness A measure of the flattening of an oblate spheroid, which is the ratio of the difference between the major and minor axes to the major axis.

obliquity of the ecliptic The acute angle between the ecliptic and the celestial equator ($23\frac{1}{2}°$).

obscuration Absorption of starlight by interstellar dust.

occultation An eclipse of a more distant celestial object by the moon or one of the planets.

ocular An eyepiece.

Oort's constants Constants that characterize the rotation of the Milky Way Galaxy in the solar neighborhood.

opacity The ability of a substance to resist the flow of radiation through it.

open cluster A loose collection of stars numbering from a few dozen up to a few hundred that lies in the galactic disk; a galactic cluster.

opposition The planetary configuration in which the earth lies between the sun and the planet; the planet thus appears to be "opposite" the sun in the sky as viewed from the earth.

optical binary (or double) A pair of stars that are along the same line of sight but are at different distances and hence are not a physical pair.

optical depth A measure of the reduction in the intensity of a beam of radiant energy as it passes through an absorbing medium.

orbit The path of one celestial object about another.

ozone An extremely reactive form of molecular oxygen composed of three oxygen atoms instead of the normal two.

parabola A curve formed by the intersection of a right circular cone and a plane parallel to the cone's surface.

parallax The apparent displacement of a nearby object relative to a background as a result of a change in the observer's position.

parsec The distance to a star that exhibits a parallax of 1 arcesecond when viewed from a separation of 1 AU; 3.1×10^{13} km; 3.26 light-years.

partial eclipse An eclipse in which the object being eclipsed is not completely obscured.

peculiar velocity The velocity of a star relative to the local standard of rest.

penumbra (shadows) A portion of an object's shadow partially illuminated by the light source.

penumbra (sunspots) The lighter outer regions of a sunspot.

penumbral eclipse An eclipse in which an object passes only through the eclipsing object's penumbra.

perfect gas law The statement that the pressure of the gas is proportional to the product of its density and temperature.

periastron The point of closest approach in the relative orbit of a binary star system.

perigee The point of closest approach to the earth of an object in earth orbit.

perihelion The point of closest approach to the sun of an object orbiting the sun.

period The interval of time required for a phenomenon or event to repeat itself.

periodic comet A comet that returns to the vicinity of the sun at more or less regular intervals.

period-luminosity relation An observed relationship between the period of light variation and the luminosity for certain types of variable stars, especially the cepheid variables.

perturbation A small-scale departure from the idealized behavior of a system, especially a two-body gravitating system, due to external forces.

phases The apparent changes of shape of an object as its illuminated surface is viewed from various angles.

photoelectric effect An effect in which electrons are emitted from a surface exposed to light.

photoelectric photometer A device that measures the intensity of radiation using the photoelectric effect.

photographic magnitudes The magnitude obtained for an object using blue-violet sensitive photographic plates.

photometry The branch of astronomy that deals with the measurement of the brightnesses of celestial objects.

photomultiplier A photoelectric cell in which the flow of electrons is amplified by successive usages of the photoelectric effect.

photon A bundle or ''particle'' of radiant energy characterized by a wavelength and an energy; the unit of electromagnetic energy.

photosphere The layer of solar or stellar atmosphere that marks the visible disk of the sun or star.

photovisual magnitude   The magnitude obtained for an object using yellow- or green-sensitive photographic plates.

plages   Bright regions in the magnetic fields surrounding sunspots visible on spectra heliograms of the sun; flocculi.

planet   One of the nine largest nonluminous objects orbiting the sun; a nonluminous object orbiting a distant star.

planetarium   An optical instrument that can project representations of the night sky and its associated phenomena onto a domed ceiling.

planetary nebula   A bright spherical nebula that surrounds a hot central star and appears as a planetary disk in a telescope.

planetesimals   One of a number of small bodies a few hundred meters across believed to have formed into protoplanets in the primeval solar system.

planetoid   A minor planet.

Planck's law   A mathematical expression of the distribution of energy with wavelength for a blackbody radiator.

plasma   A gas consisting entirely of ionized atoms.

polar axis   The axis in an equatorial telescope mounting that is aligned parallel to the earth's axis of rotation.

polarization   The alignment of the vibration planes of an electromagnetic wave into a single preferred plane.

Population I stars   Stars similar to the sun in their chemical composition.

Population II stars   Stars having significantly lower metal abundances than the sun.

position angle   The orientation of one object relative to a second nearby object.

positron   A subatomic particle having a positive charge and a mass equal to that of the electron; an antielectron.

potential energy   Stored energy that can be converted into other forms of energy.

precession   The slow, conical motion of the earth's axis of rotation due to the gravitational effects of the sun and moon on the earth's equatorial bulge.

primary minimum   The deepest drop of the light curve of an eclipsing binary system.

prime focus   The focal point of the primary mirror of a reflecting telescope.

prime meridian   The Greenwich meridian, 0° longitude.

primeval atom   In cosmological theory, the single mass from which the universe originated.

principle of equivalence   The statement that any point in space-time can be transformed into a frame of reference such that gravitational effects will disappear.

prism   A triangular piece of glass that can break up light into its component colors and create a spectrum.

prominence   A region of bright gas protruding from the solar limb.

proper motion   The apparent angular change of position of a star per year due to its intrinsic motion.

proton   A subatomic particle that carries a positive charge and is, along with the neutron, a basic constituent of atomic nuclei.

proton-proton chain   A series of thermonuclear reactions in which protons are built up into helium nuclei with an attendant conversion of mass into energy.

pulsar   A radio source, believed to be a neutron star, that emits highly regular, very short period bursts of radio emissions.

pulsating variable   A variable star whose light variations arise from successive expansions and contractions of the star.

quadrature   A planetary configuration in which the angular separation between the planet and the sun is 90° as seen from the earth.

quarter moon   Either of the two lunar phases in which the moon is 90° from the sun.

quasars   Star-like radio sources whose spectra show extremely large redshifts; quasi-stellar objects.

quiet sun   A term applied to the sun when its activity cycle is at a minimum.

radar astronomy   The branch of astronomy that deals with observation of objects by reflecting radio waves from their surfaces.

radial velocity   The component of an object's velocity that is directly toward or away from the observer.

radiant   The point on the celestial sphere from which a meteor shower appears to radiate.

radiation pressure   The pressure exerted by electromagnetic radiation on the body it strikes.

radio telescope   An instrument designed to collect and observe radio waves.

radio waves   The region of the electromagnetic spectrum that has wavelengths longer than 1 mm.

radiation   A mechanism by which energy is transported through space; the energy that is so transferred.

radioactivity   The spontaneous breakdown of slightly unstable atomic nuclei.

radius vector   The line joining two orbiting objects.

Rayleigh scattering   The process by which molecules scatter light rays.

red giant   A large, cool star of high luminosity.

Red Spot   A gigantic cyclonic storm in the upper atmosphere of Jupiter.

reddening (interstellar)   The reddening of starlight passing through interstellar dust; the dust scatters blue light more effectively than red light.

redshift   The shift to longer wavelengths of the light from remote galaxies.

reflecting telescope (reflector)   A telescope that employs a concave mirror as its principal light-gathering element.

reflection nebula   An interstellar dust cloud illuminated by stars in or near it.

refracting telescope   A telescope that employs a lens or lens system as its principal light-gathering element.

regression of nodes   The motion of an orbit's nodes due to gravitational perturbations.

relativity   A description of motion that deals with the behavior of objects moving at very high velocities or in very strong gravitational fields.

resolution   The degree to which fine details are delineated in an image.

resolving power   The ability of an instrument to observe fine detail.

retrograde motion   The apparent ''backward'' motion of a planet relative to the stars as a result of the relative motions of the earth and planet.

revolution   The orbital motion of one object about another.

right ascension   The angular distance between the hour circle of the vernal equinox point eastward to the hour circle of the object.

rille   A crevasse or trench-like depression on the lunar surface.

Roche's limit   The smallest distance at which an orbiting object can maintain itself against the tidal forces of its primary.

rotation   The spinning motion of an object about an axis passing through itself.

RR Lyrae stars Pulsating variable stars having periods of less than a day, usually found in globular star clusters.

Russell-Vogt theorem A theorem in astrophysics which states that the entire structure of a star is uniquely determined from its mass and composition.

S-type star A cool star whose spectrum is characterized by molecular bands of the heavy metal oxides such as ZrO.

saros A cycle of similar eclipses recurring about every 18 years.

satellite An object that revolves around a much larger object; a moon.

scale For an optical instrument, the angular distance of sky covered per unit distance of image formed.

Schmidt telescope An optical system in which a spherical mirror and a glass correcting plate are employed to obtain sharp images over a wide field of view.

scientific method An investigative approach in which results of experiments and observation are used to formulate hypotheses that are then tested with further experimentation.

secondary minimum The shallowest of two drops in the light curve of an eclipsing binary.

secular Nonperiodic.

secular parallax A parallax (or distance) derived from the mean motion of a set of stars that is due to the sun's motion through space.

seeing The quality of steadiness in the earth's atmosphere; unsteadiness blurs telescopic images.

seismic Related to vibrations in the outer layers of the earth's interior.

semimajor axis Half the length of an ellipse's major axis.

semiregular variable A pulsating variable star having a period that is not quite constant.

separation The angular distance between two celestial objects, especially the two members of a binary system.

Seyfert galaxy A galaxy having a bright nucleus, but not as prominent as the nuclei of N galaxies.

shell star A star surrounded by a thin, detached sphere of gas.

sidereal Of or pertaining to the stars.

sidereal day The interval of time between two successive crossings of the celestial meridian by the vernal equinox.

sidereal period The amount of time needed for one object to orbit another using the stars as a reference.

sidereal time The local hour angle of the vernal equinox.

siderite A meteorite composed of 90 percent iron, 9 percent nickel, and 1 percent other substances; an iron meteorite.

siderolite A meteorite made up of 50 percent iron and 50 percent silicates; a stony-iron meteorite.

solar antapex The point on the celestial sphere from which the sun has come.

solar apex The point on the celestial sphere toward which the sun is moving.

solar constant The amount of solar radiation striking the earth's surface per unit area per unit time; 1.4×10^6 ergs/cm^2/sec.

solar day The interval of time between two successive crossings of the celestial meridian by the sun.

solar eclipse An eclipse of the sun by the moon.

solar motion. The motion of the sun relative to the local standard of rest.

solar parallax The angle subtended by the earth's equatorial radius at a distance of 1 AU.

solar system The entire system of planets, satellites, minor planets, comets, and meteoroids that orbit the sun.

solar time The hour angle of the sun plus 12 hours.

solar wind A high-speed outward flow of gas from the sun.

solstice The points on the celestial sphere where the sun reaches its maximum angular distances north and south of the celestial equator.

south point The point of intersection between the horizon and the celestial meridian that is closest to the south celestial pole.

space motion (or velocity) The velocity of a star relative to the sun.

space-time A relativistic view of space in which time is regarded as a fourth dimension.

specific gravity The ratio of the density of an object to that of water.

spectral class (or type) The classification of a star based on the appearance of its line spectrum.

spectral sequence An arrangement of spectral classes in order of increasing or decreasing temperature.

spectrograph An instrument used to record the spectrum of an object.

spectroheliogram A photograph of the sun taken in monochromatic light, usually at the wavelength of a strong hydrogen or calcium absorption line.

spectroscopic binary A binary star whose orbital motion manifests itself as a variable radial velocity.

spectroscopic parallax   The parallax (or distance) obtained for a star by comparing its apparent magnitude with its absolute magnitude as deduced from the star's spectral characteristics.

spectrum   The radiant energy from an object spread out into its component wavelengths by some dispersive device such as a prism or grating.

spectrum variable   A star whose spectral characteristics vary in time.

speed   The magnitude of a velocity.

spherical aberration   A defect in an optical system in which light rays for an axis striking near the edges of the lens or mirror come to a different focus than light rays striking near its center.

spicule   A narrow, upward jet of material in the solar chromosphere.

spiral arms   Curved cylindrical regions of gas, dust, and stars that wind outward in a plane from the nucleus of a spiral galaxy.

spiral galaxy   A galaxy that has spiral arms.

sporadic meteor   A random meteor not associated with a meteor shower.

spring tide   The highest tides that can occur in a given month; spring tides occur around the time of new moon or full moon.

standard time   The local mean solar time of a standard meridian that is extended to large areas on both sides of the meridian for convenient timekeeping; a variation of zone time.

star   A self-luminous sphere of gas.

star cluster   An assemblage of stars held together by the mutual gravity of its members.

statistical parallax   A parallax (or distance) derived from the mean motion of a set of stars that is independent of the solar motion.

steady-state theory   In cosmology, an outdated theory in which the density of matter in an expanding universe is kept constant through the continuous creation of matter.

Stefan's law   The statement that the total amount of radiation emitted from a blackbody per unit area of surface is proportional to the fourth power of the object's temperature.

stellar evolution   The life cycle of a star, including all of the physical changes that occur during that cycle.

stellar model   A theoretical calculation of the interior conditions for a star.

stellar parallax   The angle subtended by 1 AU at the distance to a given star.

stones   A type of meteorite composed mostly of silicates; an aerolite.

stony-iron A type of meteorite composed of about 50 percent silicates and 50 percent iron and nickel; a siderolite.

stratosphere The layer of the earth's atmosphere lying above the troposphere and below the ionosphere.

Strömgren sphere A region of ionized gas in space surrounding a hot star.

subdwarf A star that lies below the main sequence on the H-R diagram.

subgiant A star that lies between the main sequence and the giant branch on the H-R diagram.

summer solstice The point on the celestial sphere at which the sun attains its largest positive declination.

sunspot A dark, cool region of high magnetic field intensity in the solar photosphere.

supergiant A star of extremely large size and luminosity.

superior conjunction A conjunction of an inferior planet and the sun in which the planet is on the far side of the sun.

superior planet A planet whose orbit lies outside that of the earth.

supernova A catastrophic outburst of a star in which large amounts of its mass are ejected into space; its energy output can increase by millions of times.

surface gravity The acceleration due to gravity at the surface of an object.

synchrotron radiation Radiation emitted by high-speed charged particles, especially electrons, as they are accelerated in a magnetic field.

synodic month The interval of time between successive appearances of the same lunar phase.

synodic period The amount of time it takes for a celestial configuration to repeat itself.

syzygy A configuration of the moon when the earth, sun, and moon are in a line; a new or full moon.

T Tauri stars Irregular variable stars believed to be in a phase of their evolution just prior to becoming main-sequence stars.

tangential velocity The component of an object's velocity that is perpendicular to the observer's line of sight.

tektites Rounded glassy objects believed by some scientists to be of extraterrestrial origin.

telescope An optical device that enhances the astronomer's view of the heavens.

telluric Terrestrial in origin.

temperature   A measure of the internal energy of a body.

terminator   The line between the lit and unlit positions of a reflecting body.

terrestrial planets   The small, compact planets nearest the sun, including Mercury, Venus, Earth, and Mars; Pluto is sometimes included in this group.

thermal energy   The energy associated with the motions of atoms or molecules in a given object or substance.

thermal equilibrium   A state of balance between the amount of heat flowing into and out of a given system.

thermocouple   An electrical device used by astronomers to measure the intensity of infrared radiation.

thermodynamics   The branch of physics that deals with the properties of heat and heat transfer.

thermonuclear   Of or pertaining to a high-temperature fusion reaction.

tidal force   The stretching of an object due to the gravitational forces of a nearby object.

total eclipse   An eclipse in which the light from one object is completely blotted out by a second object.

totality   The interval of time during a total eclipse in which the light from one object is completely blotted out by another solid body.

transit (astrometry) The passage of a celestial object across the celestial meridian; an instrument for measuring the exact time and altitudes of transits.

transit (solar system)   The passage of a small body in the solar system across the disk of a larger one.

transition   A change of the electronic configuration in an atom.

triangulation   A method of distance determination by which the distance to an inaccessible point is obtained by computing the elements of a triangle involving the point.

trigonometry   The branch of mathematics that deals with analytical solutions of triangles.

triple-alpha (3-α) process   A sequence of nuclear reactions in which three helium nuclei are fused into a carbon nucleus.

Trojan asteroids   The set of minor planets that move about the sun in Jupiter's orbit 60° ahead of and 60° behind the planet itself.

Tropic of Cancer   The parallel of latitude on the earth over which the sun stands at the summer solstice; 23½°N latitude.

Tropic of Capricorn   The parallel of latitude on the earth over which the sun stands at the summer solstice; 23½°N latitude.

tropical year   The interval of time required for the sun to make successive passages through the vernal equinox point.

troposphere   The lowest, densest layer of the earth's atmosphere.

turbulence   The irregular, random motions in a gas or liquid.

U,B,V system   A photometric system in which magnitudes of objects are measured in the ultraviolet (U), blue (B), and yellow-green (V) regions of the electromagnetic spectrum.

ultraviolet   That part of the electromagnetic spectrum lying roughly between 1000 Å and 4000 Å.

umbra (shadows)   The completely dark central portion of an object's shadow.

umbra (sunspots)   The dark central portions of a sunspot.

universal time   The local mean time of the Greenwich meridian.

universality   An assumption that the laws of nature are invariant in time or with one's location in the universe.

universe   The sum total of all that we see in the heavens.

UV Ceti stars   Red flare stars.

Van Allen belts   A set of doughnut-shaped regions about the earth in which high-energy charged particles have been trapped by the earth's magnetic field.

variable star   A star whose spectral characteristics or brightness varies with time.

variation of latitude   Small-scale changes in the latitudes of places on the earth due to a change in the orientation of the earth's axis of rotation relative to the surface of the earth.

velocity   The time rate of change of an object's position in some specified direction.

velocity of escape   See escape velocity.

vernal equinox   The point on the celestial sphere where the sun crosses the celestial equator passing from south to north; the day on which this passage occurs.

vertical circle   Any great circle passing through the observer's zenith.

virial theorem   A relationship between the total potential and kinetic energy of a given system.

visual binary (star)   A binary star in which both components can be resolved optically.

visual magnitude   The magnitude of an object in the yellow-green or ''visual'' region of the electromagnetic spectrum.

volume   A measure of the total space occupied by a given object.

Vulcan   A planet hypothesized in the nineteenth century to be between Mercury and the sun.

W Ursae Majoris star   An eclipsing binary whose components are nearly in contact with one another.

walled plain   A large lunar crater with small outside walls.

wandering of the poles   The change in the orientation of the earth's axis of rotation relative to its surface that is responsible for the variation of latitude effect.

waning moon   The set of lunar phases between full and new moon when the amount of illuminated surface is decreasing in time.

wavelength   The distance between successive crests (or troughs).

waxing moon   The set of lunar phases between new and full moon when the amount of illuminated surface is increasing in time.

weight   The total gravitational force exerted on a given mass by a given object.

west point   The point on the horizon 90° to the left of the north point.

white dwarf   A final stage of a star's evolution in which it has collapsed to an object roughly the size of the earth and can generate no more nuclear energy.

Widmanstätten figures   A unique crystalline structure that can be seen on the face of a cut and polished iron meteorite.

Wien's law   The relationship that states that for any blackbody, the product of the temperature and the wavelength at which the largest energy output occurs is a constant.

winter solstice   The point on the celestial sphere where the sun is farthest south of the celestial equator.

Wolf-Rayet star   A star whose spectrum is characterized by broad emission bands of oxygen and nitrogen that arise from ejected gaseous shells.

X rays   Radiation from that part of the electromagnetic spectrum lying between 1 Å and 1000 Å.

year   The time required for the earth to complete one orbit of the sun.

Zeeman effect   The splitting of the spectral lines of an atom due to the effect of a magnetic field on the atom's electronic energy levels.

zenith   The point on the celestial sphere directly overhead.

zero-age main sequence   The sequence of positions on the H-R diagram reached by protostars of various masses at the onset of nuclear fusion processes in their cores.

zodiac   An imaginary band on the celestial sphere approximately 16° wide which is centered on the ecliptic and within which the sun, moon, and all of the planets (except Pluto) move.

zodiacal light   A faint glow of light along the zodiac believed to be due to sunlight reflected and scattered by interplanetary dust.

zone of avoidance   An irregular band roughly coincident with the outlines of the Milky Way within which very few exterior galaxies can be seen.

zone time   The time kept in a strict 15°-wide longitude zone that is equal to the local mean time for the zone's central meridian.

Index